OECD Studies on Water

Financing Water Supply, Sanitation and Flood Protection

CHALLENGES IN EU MEMBER STATES AND POLICY OPTIONS

This document, as well as any data and map included herein, are without prejudice to the status of or sovereignty over any territory, to the delimitation of international frontiers and boundaries and to the name of any territory, city or area.

The statistical data for Israel are supplied by and under the responsibility of the relevant Israeli authorities. The use of such data by the OECD is without prejudice to the status of the Golan Heights, East Jerusalem and Israeli settlements in the West Bank under the terms of international law.

The present publication presents time series, which end before the United Kingdom's withdrawal from the European Union on 1 February 2020. The EU aggregate presented here therefore refers to the EU including the UK. In addition to being included in the EU aggregate, the UK also features in relevant tables and figures, when there is a breakdown of the data by country.

Note by Turkey
The information in this document with reference to "Cyprus" relates to the southern part of the Island. There is no single authority representing both Turkish and Greek Cypriot people on the Island. Turkey recognises the Turkish Republic of Northern Cyprus (TRNC). Until a lasting and equitable solution is found within the context of the United Nations, Turkey shall preserve its position concerning the "Cyprus issue".

Note by all the European Union Member States of the OECD and the European Union
The Republic of Cyprus is recognised by all members of the United Nations with the exception of Turkey. The information in this document relates to the area under the effective control of the Government of the Republic of Cyprus.

Please cite this publication as:
OECD (2020), *Financing Water Supply, Sanitation and Flood Protection: Challenges in EU Member States and Policy Options*, OECD Studies on Water, OECD Publishing, Paris, *https://doi.org/10.1787/6893cdac-en*.

ISBN 978-92-64-67888-0 (print)
ISBN 978-92-64-66438-8 (pdf)

OECD Studies on Water
ISSN 2224-5073 (print)
ISSN 2224-5081 (online)

Photo credits: Cover © SnvvSnvvSnvv/Shutterstock.com; © Sebestyen Balint/Shutterstock.com

Corrigenda to publications may be found on line at: *www.oecd.org/about/publishing/corrigenda.htm*.
© OECD 2020

The use of this work, whether digital or print, is governed by the Terms and Conditions to be found at *http://www.oecd.org/termsandconditions*.

Foreword

In the context of the European Green Deal - the roadmap for making the EU's economy sustainable - and the recovery from the economic and social crises triggered by the coronavirus, investments in water can contribute to sustainable growth and to building resilience for communities to a range of water and health risks. More specifically, compliance with the Water Framework, Drinking Water, the Urban Wastewater Treatment and the Flood Directives contributes to the health of European citizens, protects water bodies and ecosystems, and enhances the resilience of our communities and economies.

The recent Fitness Check of the Water Framework Directive and the Floods Directive and the evaluation of the Urban Wastewater Treatment Directive confirm that these directives are still relevant after several decades and create value for money. While the economic case for achieving compliance with the water *acquis* is robust, several countries still find it difficult to mobilise the required level of investment.

In that context, the European Commission and the OECD joined forces to assess the scale of investment still required to achieve compliance with the EU water *acquis,* to better understand financing capacities at country level and to explore options to bridge the financing gap. The European Commission and member states can reflect on the Recommendations derived from this analysis to inform their activities to reach compliance, at better cost for communities.

This report illustrates the unique combination of data, analytical skills, policy insights and convening power that the European Commission and the OECD can leverage together. It also illustrates the benefit of a dialogue between our respective institutions and constituencies. We learned a lot from this process and the interactions with national and local authorities, civil society organisations, corporations and financiers. We are grateful to the countries, professionals and experts who contributed throughout the process.

This collaboration is already having concrete impacts. First, it informs regulatory reforms the European Commission seeks to undertake. Second, it inspires reform agendas in several EU member states, with support from the European Commission Structural Reform Support. It will also inform further analyses and policy discussions in European institutions and OECD bodies, including through a future regional meeting of the Roundtable on Financing Water co-convened by the OECD and the European Investment Bank. This collaboration continues, as the European Commission and the OECD facilitate a series of thematic workshops to support the implementation of the policy recommendations in this report.

This report is well worth a read. We trust it can inspire result-oriented action to the benefit of our member states and beyond. Its messages go beyond the water community and are relevant for governments and partners committed to deliver water-wise, sustainable and resilient development, at international, national or local scales.

Daniel Calleja
Director General, DG Environment
European Commission

Rodolfo Lacy
Director of the Environment Directorate
Organisation for Economic Co-operation and Development (OECD)

Acknowledgments

This document reports on a joint project by the European Commission DG Environment and the OECD on financing needs and capacities of EU member states. The OECD team was coordinated by Xavier Leflaive. Nathalie Cliquot, Lisa Danielson, Kathleen Dominique, Tatiana Efimova, Raphaël Jachnik, Hannah Leckie and Harry Smythe contributed substantively. Simon Buckle, Head of the CBW Division, provided guidance. Ines Reale and Anna Rourke provided valuable assistance.

The OECD benefitted from substantial inputs and active cooperation from two teams of international experts: Acteon, led by Pierre Strosser, with Rianne van Duinen and Gloria De Paoli; and EcoFutures, led by Richard Ashley, with Bruce Horton. Bernard Barraqué contributed on selected topics.

Experts contributed to country-specific analyses: Irina Ribarova (Bulgaria), Vesna Simic (Croatia), Ayis Iacovides (Cyprus), Dionysis Assimacopoulos (Greece), Kristina Veidemane (Latvia), Daiva Seméniené (Lithuania), Krzysztof Berbeka (Poland), Elena Fatulova (Slovakia), Gonzalo Delacámara (Spain).

Preliminary analyses were discussed at a workshop co-convened by the OECD and the European Commission, in Brussels, on 18 May 2018. Policy recommendations were discussed at a workshop co-convened by the OECD and the European Commission, in Brussels, on 6 December 2019.

National workshops were organised in 9 countries, gathering senior officials and a wide range of stakeholders, to fine-tune the understanding of the financing challenges and explore policy options. The authorities of the countries are gratefully acknowledged: Bulgaria, Croatia, Cyprus, Lithuania, Poland, Romania, Slovakia, Slovenia, and Spain.

The European Commission provided constructive guidance throughout the process: Michel Sponar and Hans Stielstra, with Lie Heymans, Trudy Higgins and Anna Marczak. Nele-Frederike Rosenstock managed the project beautifully on both substance and process. They are gratefully acknowledged.

The project would not have been possible without the financial support from the European Commission - DG Environment. It also benefitted from and contributed to the work of the *Roundtable on Financing Water*, a joint initiative by the OECD, the government of the Netherlands, the World Water Council and the World Bank.

Table of contents

Foreword — 3

Acknowledgments — 5

Executive summary — 10

Background and process — 17
    Notes — 18

1 Framing the challenge — 19
    1.1. The benefits of investing in water — 20
    1.2. Ambition and scope of the project — 23
    1.3. Drivers of investment needs — 24
    1.4. Emerging challenges — 31
    References — 43
    Notes — 44

2 The state of play — 45
    2.1. Water supply and sanitation — 46
    2.2. Flood protection — 57
    References — 61
    Notes — 61

3 Projected investment needs across member states — 63
    3.1. Water supply and sanitation — 64
    3.2. Flood protection — 74
    3.3. Investment needs under the Water Framework Directive — 78
    References — 84
    Notes — 85

4 The capacity to finance projected investment needs across member states — 87
    4.1. Financing capacity — 88
    4.2. Preliminary conclusions — 93
    References — 95

5 Selected options to address financing challenges — 97
    5.1. Options to make the best use of existing assets and financial resources — 99
    5.2. Options to minimise future financing needs — 105

5.3. Financing as part of flood risk mitigation strategies	121
References	129
Notes	131

Annex A. Costs of addressing emerging challenges in wastewater collection and treatment	133
Annex B. Data and method	135
Annex C. Data supporting the results on projected coastal flood risk investment needs	136
Annex D. Projections by EurEau on costs of compliance with DWD and UWWTD	139
Annex E. Assessment of RBIs and RFIs to finance flood protection	141

Tables

Table 1. Policy Recommendations to meet water-related financing needs in Europe	15
Table 1.1. Rates of expenditure for "best estimate" scenarios	30
Table 1.2. Projected impacts of climate change on water across Europe	33
Table 1.3. Household pharmaceutical collection and disposal programmes, select OECD countries	40
Table 3.1. Converging towards 5% IAS per country	70
Table 3.2. Country clusters based on projected exposure to riverine floods	76
Table 3.3. Projected coastal flood risk investment needs	78
Table 3.4. Examples of investment options to improve water quality	81
Table 4.1. Sovereign credit rating	91
Table 4.2. Member states' capacity to finance projected investment needs for WSS	94
Table 5.1. Policy Recommendations to meet water-related financing needs in Europe	98
Table 5.2. Benefits of nature-based solutions versus traditional engineered 'grey' infrastructure for urban water management	107
Table 5.3. Overall comparison of NBS and grey infrastructure for flood protection	111
Table 5.4. Technologies and innovations for water systems	114
Table 5.5. Selected existing economic instruments in Europe	123
Table 5.6. Selected existing insurance schemes in the EU	125
Table 5.7. Innovative financing mechanisms for flood protection	126
Table 5.8. The performance of economic policy instruments as part of risk mitigation strategies	127
Table A A.1. Cost estimates for managing emerging and associated pollutants	133
Table A C.1. Data supporting the results on projected coastal flood risk investment needs	136
Table A D.1. Investment needs in water infrastructure reported by member states	139
Table A D.2. Expenditure needs to renew existing infrastructure reported by member states	140
Table A E.1. Assessment of RBIS and RFIs to finance flood protection	141

Figures

Figure 1.1. Relative economic impacts of water insecurity	23
Figure 1.2. Share of population residing in urban areas (%)	26
Figure 1.3. Stylised sequence of investments in water supply, sanitation and flood protection	28
Figure 1.4. Total investment needs by 2040 for CEC treatment – extrapolation of the Swiss approach	38
Figure 2.1. Rate of asset renewal for water supply & Rate of asset renewal for sanitation	47
Figure 2.2. Estimated annual expenditures for water supply and sanitation for the EU-28	48
Figure 2.3. Estimated annual expenditures for water supply and sanitation per member state	48
Figure 2.4. Estimated expenditures per capita for water supply and sanitation in EU-28	49

Figure 2.5. Estimated annual expenditures per capita for water supply and sanitation per member state 49
Figure 2.6. Estimated expenditures per capita and as % of GDP 50
Figure 2.7. Sources of finance for water supply and sanitation services for the EU-28 51
Figure 2.8. Sources of finance for water supply and sanitation services per member state 51
Figure 2.9. Share of EU funding in estimated total expenditures for water supply and sanitation for the EU-28 52
Figure 2.10. Share of EU funding in estimated total expenditures for water supply and sanitation per member state 53
Figure 2.11. Share of debt in estimated total expenditures for water supply and sanitation for the EU-28 53
Figure 2.12. Share of debt in estimated total expenditures for water supply and sanitation per member state 55
Figure 2.13. Share of water supply and sanitation expenditures in households' disposable income 56
Figure 2.14. Estimated public budget expenditure for flood protection for the EU-28 58
Figure 2.15. Estimated public budget expenditure for flood protection per member state 58
Figure 2.16. Share of EU funding in public budget expenditures for flood protection for the EU-28 59
Figure 2.17. Share of EU funding in public budget expenditures for flood protection per member state 60
Figure 3.1. Additional expenditures by 2030 for water supply and sanitation Baseline + Business as usual scenario 65
Figure 3.2. Investment needs for water use efficiency per member state 67
Figure 3.3. Additional expenditures by 2030 for water supply - Compliance & efficiency scenario 69
Figure 3.4. Additional expenditure by 2030 for sanitation – Compliance scenario 71
Figure 3.5. Total cumulative additional expenditures by 2030 for water supply and sanitation 72
Figure 3.6. Per capita cumulative additional expenditures by 2030 73
Figure 3.7. Per Annum additional expenditures by 2030 74
Figure 3.8. Total growth factors for river flood risk expenditure by 2030 75
Figure 3.9. Cost comparison of options to reduce a nitrogen loading, Chesapeake Bay Watershed, United States 82
Figure 4.1. Projected affordability issues – constant households disposable income 89
Figure 4.2. Affordability of water supply and sanitation compounded by risk of poverty 90
Figure 4.3. Ability to increase public spending based on raising taxes or borrowing 91
Figure 4.4. Domestic credit to private sector 92

Boxes

Box 1. Pending Issues – The case of pharmaceutical residues in freshwater 13
Box 1.1. Water security, defined 21
Box 1.2. Relative economic impacts of water insecurity 22
Box 1.3. Experts' view on future drivers of water infrastructure investment needs 25
Box 1.4. Water mains and sewerage infrastructure renewal in England and Wales 30
Box 1.5. Managing combined sewer overflows 34
Box 1.6. Efforts to adapt water management to climate change in OECD countries 35
Box 1.7. Addressing pharmaceutical residues in freshwater. The Swiss approach 37
Box 1.8. A proposal for an EPR Scheme to recover costs of advanced wastewater treatment plant upgrades, Germany 41
Box 1.9. Policy responses to CECs: A state of flux 42
Box 2.1. Financing needs and capacities for WSS in Sweden 54
Box 3.1. Towards accurate knowledge of water supply and sanitation assets 66
Box 3.2. An arbitrary threshold for leakage reduction 67
Box 3.3. Additional expenditures to converge towards 5% IAS per country 70
Box 3.4. A case of costly historical pollution: Flix Reservoir, Ebro Basin, Spain 80
Box 3.5. A voluntary agreement to stimulate investment in the protection of water bodies, in New Zealand 83
Box 5.1. Nature-based solutions to manage stormwater in Philadelphia, USA 108
Box 5.2. Land value capture – a suite of tools to finance water-related investments 117
Box 5.3. European experience with bending sources of finance for water-related investment 119

Follow OECD Publications on:

 http://twitter.com/OECD_Pubs

 http://www.facebook.com/OECDPublications

 http://www.linkedin.com/groups/OECD-Publications-4645871

 http://www.youtube.com/oecdilibrary

 http://www.oecd.org/oecddirect/

Executive summary

Background and objective

Member States of the European Union share the same level of ambition for water policies and management, set out by the Water Framework Directive (2000/60/EC): a series of technical directives contribute to achieving those ambitions. Three deserve particular attention: the Urban Waste Water Treatment Directive (UWWTD; 91/271/EEC); the Drinking Water Directive (DWD; 98/83/EC); and the Floods Directive (FD; 2007/60/EC).

Compliance with these technical directives contributes to achieving the ambition of the Water Framework Directive. More specifically, it contributes to a series of benefits for communities and member states. Compliance with the DWD contributes to inclusive health and hygiene. Compliance with the UWWTD contributes to minimising the load of pollutants in freshwater streams and the sea. Since the adoption of the UWWTD in 1991, the load of Biochemical Oxygen Demand, nitrates and phosphorus in treated urban waste water have decreased by 61%, 32% and 44% respectively, contributing to improved quality of surface water and coastal waters (European Commission, 2019, Evaluation of the UWWTD). This translates into minimising treatment costs downstream, healthy freshwater ecosystems, and improved bathing water quality, among other direct and indirect benefits. Implementation of the FD has supported a shift from policies based on flood defence, towards flood risk assessment, and is a potential template for best practices in disaster management (European Commission, 2019, Water Fitness check).

Still, several countries do not comply with the three technical directives. In the case of the revised DWD, some vulnerable groups or marginalised communities may not have access to safe drinking water. As regards urban wastewater collection and treatment, the UWWTD mandates secondary level of treatment, which remains an objective in some territories. The UWWTD also requests more stringent treatment in sensitive areas. Several countries, especially in rural communities, rely on Individual and other Appropriate sanitation Systems (IAS; for instance, sceptic tanks), and it is not always clear how the performance of such systems is monitored and compliance with the UWWTD is enforced. Another area of concern are combined sewer overflows and urban runoff. In times of climate change and recurring heavy rainfall events, the pollution from these sources becomes increasingly important to address.

Drinking water, urban wastewater collection and treatment, and flood protection are affected by emerging issues, which may put additional pressure on vital infrastructure and services. For instance, in the context of the evaluation of the UWWTD, the European Commission has identified issues such as contaminants of emerging concerns (CECs; essentially pharmaceutical residues or microplastics in freshwater), combined sewer overflows, small agglomerations and IAS, and sludge management as issues that need to be addressed to ensure that wastewater collection and treatment contribute to the objectives of the WFD and related priorities across Europe, now and in the future.

Limited availability of and access to finance are often mentioned by member states to explain distance to compliance or raise concern about the capacity to comply with future regulations on water supply and sanitation. Sufficient finance is needed to cover the investment needs for the three technical directives, to operate and maintain infrastructure and ensure good service and performance, and to respond to emerging challenges in the future.

The OECD and the European Commission joined forces to i) document investment needs member states face to comply (and remain compliant) with the DWD, UWWTD and FD, now and in the future, and to ii) assess financing capacities and characterise financing challenges more precisely. This analysis can support discussions on the options countries may wish to consider to close the financing gap. It can also help position and calibrate the support the European Commission can provide to member states to ensure compliance with the three directives at least cost for the community.

Method and data

Projections of future investment needs derive from a baseline of current expenditures (based on best-available and comparable data) and the influence of several drivers of investment needs. Three scenarios are considered:

- Business as usual for water supply and sanitation services. This scenario projects the same level of effort, with no new policies. Projections are driven by urban population growth (see below the discussion on drivers). The projections reflect the current level of effort: they do not consider the potential delay or backlog of investment and the state of existing infrastructures. Potential under-spending in the operation, maintenance and renewal of existing assets will continue under this scenario, potentially leading to significant additional investment needs in the longer term.
- For water supply: projections to achieve compliance, efficiency and access. Most EU member states already comply with, or are close to complying with, the Drinking Water Directive (DWD). It is anticipated that, even when member states comply with the revised DWD, countries will need to invest in water efficiency and minimise non-revenue water (including leakage). In addition, countries will have to ensure that vulnerable groups have access to safe water.
- For sanitation: projections to achieve compliance. Several EU member states do not fully comply with the Urban Wastewater Treatment Directive (UWWTD). The extent of compliance varies across EU member states and has been considered the main driver for additional investment in this domain.

The current level of efforts in flood protection was not monetised. Only a few countries monitor financial flows for flood protection, usually the ones who can be expected to spend the most (Austria, the Netherlands). It was not possible to extrapolate based on available data. Therefore, projections on investment needs for flood protection are based on changes in the exposure to flood risks.

Emerging challenges, which could not be monetised, are discussed qualitatively. These include climate change and contaminants of emerging concern (e.g. focused primarily on pharmaceuticals for the purpose of this analysis). A rough estimate of investment needs to address contaminants of emerging concern is presented at an aggregate level, using costs measured in Switzerland.

Obviously, options to minimise financing needs exist and will be considered by most countries. This is the case, notably, of distributed systems or nature-based solutions for sanitation and for flood protection. How these options will materialise and affect investment needs in each member state remains highly uncertain. Therefore, such options are discussed in the report, but not reflected in the monetised projections.

The method and data used to support the baseline and the projections are synthesised in Annex B of the report. They are described in more detail in a separate methodological note.

Projections of investment needs to comply with the DWD and the UWWTD

Baseline estimates point out to an annual average expenditure of EUR 100 billion across the 28 EU member states, with the lion's share attributable to EU15 (Germany, France, United Kingdom and Italy in particular). The aggregate figure masks huge variations. Eight EU member states spend less than EUR 100 per capita per year on water supply and sanitation services. At the other end of the spectrum, six countries spend more than EUR 250.

Countries vary according to the level of efforts allocated to water supply and sanitation. Slovenia, the Czech Republic or Cyprus allocate a larger share of their GDP to water supply and sanitation than Estonia, Denmark, Sweden or Finland. This reflects the costs of the service and local conditions, and level of effort in the investment and operation of the service. This may also reflect the level of efficiency of expenditure programmes, where comparatively high levels of investment do not narrow the distance to compliance.

Looking ahead, expenditures for water supply and sanitation need to increase significantly, if countries want to comply with the DW and UWWT directives and to enhance the efficiency of water supply systems. Total cumulative additional expenditures by 2030 for water supply and sanitation amounts to EUR 289 billion for the 28 member states. Sanitation represents the lion's share of the total additional expenditures, particularly in Italy, Romania and Spain and - at lower levels – in Bulgaria, Croatia, Portugal and Slovakia. In these countries, urban population growth plays a minor part (sometimes nil) in projected future expenditures for water supply and sanitation, which are mainly driven by the need to enhance efficiency in water supply and/or compliance with the UWWTD.

A telling indicator is to compare the additional expenditures for water supply and sanitation with the current level of expenditures as captured by the baseline, on a country basis. According to the projections, all countries but Germany will need to increase annual expenditures for water supply and sanitation by more than 25% in order to comply with the directives. At the higher end, Romania and Bulgaria need to double (or more) the current level of expenditures. At the lower end of the spectrum, Cyprus, the Czech Republic, France, Germany, the Netherlands and Slovenia are projected to face comparatively minor needs for increase (by less than 1/3). This is likely to reflect different situations, including high levels of expenditures and good anticipation of future needs, significant catch-up in recent decades (Czech Republic), or underestimates of future needs, possibly driven by overreliance on IAS (Slovenia).

One pervasive challenge across member states remains financing to operate, maintain and renew existing assets. The rate of asset renewal is not known with accuracy. When it is documented, it is usually below a rate that would reflect the life expectancy of assets, suggesting that renewal efforts need to step up urgently, to avoid rapid decay of built infrastructures and degradation of service quality.

Financing capacities for water supply and wastewater collection and treatment

The OECD has identified three "ultimate" sources of finance for water supply and sanitation expenditures: revenues from water tariffs, taxes, and transfers from the international community (in Europe, essentially EU funds or to a lesser extent, concessional finance): the 3Ts. Other sources of finance (debt or equity) can be mobilised to cover the upfront costs of investments, but will need to be repaid, through a combination of the 3Ts. This rationale can be refined and characterised further, but it provides a robust heuristic to characterise financing options.

Financing capacities reflect the room for manoeuvre available to countries have with 3Ts. EU member states vary according to the ultimate source of finance mobilised for water supply and sanitation. Some rely essentially on water tariffs (Denmark, England and Wales) while others shift the burden to taxpayers (Ireland is the best example).

In some countries, public budgets allocated to water supply and sanitation heavily rely on EU funding. This is not sustainable as EU funds available for water supply and sanitation will decline over time. Therefore, EU member states need to consider more systematic reliance on domestic sources of finance to cover projected financing needs to comply with the DWD and UWWTD now and in the future.

It is difficult to assess the capacity to increase levels of public budgets allocated to water supply and sanitation, per country. Ultimately, this remains a political decision, and involves arbitrage between policy priorities. Still, macro-economic conditions and constraints can indicate room for manoeuvre to increase public spending at an aggregate level (both national and local, acknowledging that, in several countries a large share of public spending for water supply and sanitation originates in local budgets). For several countries, the current level of public debt and/or the sovereign credit rating raise concern about the capacity to allocate more public funding to expenditures related to water supply and sanitation.

With the exception of Ireland, revenues from tariffs are considered a reliable source of finance to cover at least some of the costs of water supply and sanitation services. The points above suggest that this may be even more so in the coming decades. The question then is about the room for manoeuvre to increase tariffs for water supply and sanitation services.

While affordability constraints are mentioned to justify tariffs below cost recovery levels, robust data shows that in 24 EU member states, more than 95% of the population could pay more for water supply and sanitation without facing an affordability issue (considered as a situation when households spend more than 3-5% of their disposable income on water supply and sanitation). In those countries, targeted social measures are more effective than cheap water to enhance the financial sustainability of water services while addressing the social consequences of higher tariffs.

> **Box 1. Pending Issues – The case of pharmaceutical residues in freshwater**
>
> As previously alluded to, projections do not consider a series of pending issues, most notably contaminants of emerging concern (CECs), or climate change and related issues such as combined sewer overflows. More work is needed to characterise additional pressures from these drivers, and to understand the financial implications.
>
> The report sheds some light on options to address CECs – more specifically pharmaceutical residues - in wastewater. While the presence of pharmaceutical residues can be traced in the environment, the potential adverse consequences for ecosystems, biodiversity and human health remain uncertain. Advances in analytical methods and risk assessment provide opportunities to build a policy-relevant knowledge base. Switzerland is the first country to tackle the CECs challenge at the national level. It does so through a systematic approach, which comes at a cost.
>
> The OECD identifies five strategies based on proactive policies that can cost-effectively manage pharmaceuticals for the protection of water quality and freshwater ecosystems (for more information, see OECD (2019), *Pharmaceutical Residues in Freshwater: Hazards and Policy Responses*, OECD Studies on Water, OECD Publishing, Paris, https://doi.org/10.1787/c936f42d-en). Different financing mechanisms can be considered to cover and allocate costs. Switzerland combines additional revenues from tariffs with subsidies from national budget. Other mechanisms could be considered (such as extended producers' responsibility) to minimise costs and allocate them in a fair and equitable manner.

These considerations provide a rationale to rank EU member states according to the severity of the financing challenge they face to comply with the DWD and UWWTD, now and in the future, considering both the additional level of effort required and financing capacities. Selected clusters include:

- Romania and Bulgaria face severe financing challenges as the projected additional level of effort is very high and room for manoeuvre for financing appears limited.

- Slovakia and Estonia may face similar levels of effort in the future but Estonia is better placed to cover them, as public finance looks less strained, should it need to be mobilised.
- Latvia, Poland and Portugal face similar levels of efforts in the future, but have distinct capacities in place to cover them. Affordability issues are relatively less severe in Portugal.
- The ranking of Greece and Slovenia begs questions. The additional level of effort reported by countries is probably underestimated, reflecting excessive reliance on IAS. A reassessment of additional financing needs would translate into severe challenges, as financing capacities are limited for both countries.
- The Netherlands and Germany are in privileged situations, as the additional level of efforts required is comparatively limited and financing capacities are strong.

Financing future flood protection

It was not possible to establish a robust baseline of current expenditures, as flood protection does not correspond to a sector or subsector in any international statistical standards/ international classifications. Further, survey data reported by member states are very patchy. The FD mandates the development of Flood Risk Management Plans. However, countries vary in their capacity to draft relevant planning documents and implement (and finance) them. Reported cost data for the FD in the Member State compliance assessment reports shows high variability: as an illustration, the Fitness Check calculated average capital costs per inhabitant, and those vary from EUR 0.2 in Estonia, to EUR 261 in Slovenia.

In that context, it was only possible to project additional exposure and vulnerability of countries to flood risks, taking into account annual expected affected population, urban damage, and GDP (defined as total growth factors). This was quantified for riverine floods (using WRI data) and qualitatively discussed for coastal floods. Urban floods were considered as an emerging challenge, essentially because they are not properly documented or monitored in existing data sets.

Countries can be clustered into four different categories, reflecting different perspectives on future exposure to riverine floods:

- Countries affected by the highest total growth factors (Austria, Luxembourg, the Netherlands). The increase in total growth factors is driven by climate change, indicating that urban assets, GDP and population will be increasingly exposed to flooding in the future compared to the current situation.
- Countries affected by moderate growth factors (Belgium, Czech Republic, Denmark, France, Germany, Hungary, Ireland, Poland, Romania, Slovakia, Sweden, the UK). In some of these countries, the impact of climate change is relatively low and more or less equal to the contribution of socio-economic developments in the explanation of future increases in flood risk.
- Countries benefitting from lower exposure of population (Bulgaria, Croatia, Estonia, Latvia, and Lithuania). In contrast with other member states, socio-economic developments – not changes in the climate - have a relatively large contribution to a future increase in flood risk in these countries.
- Countries benefitting from low or negative growth factors (Cyprus, Greece, Malta, Portugal, Spain). In general, these countries have limited exposure to river flood risk due to their arid or semi-arid climate (even though some catchments can be exposed to flooding during winter).

To date, flood protection in Europe has been largely financed through public grants. Alternative instruments are available to finance both investments in flood protection and the provision of financial protection in case of flood events. First, economic instruments provide a monetary/economic incentive promoting efficient flood risk management and risk reduction; they can either be administered by the government or by private entities. Second, risk financing instruments (RFIs) are pre-disaster arrangements coming into play in a post-disaster phase. They include insurance, weather derivatives and catastrophe bonds. Because they indirectly incentivise behaviour and increase the uptake and efficiency of risk-reduction measures, such instruments should be used as actual policy instruments to manage risk mitigation, together with other risk mitigation measures such as regulatory, research, and development measures. Different instruments can be combined in risk mitigation strategies to get the most out of each of them; the best mix of instruments will need to be assessed on a case-by-case basis.

Policy options member states may wish to consider

The assessment of financing challenges provides a robust basis to explore policy recommendations that can help meet financing needs. Policy recommendations that cut across the areas covered in the three directives are clustered around three sets of mutually reinforcing categories.

Table 1. Policy Recommendations to meet water-related financing needs in Europe

Make the best use of existing assets and financial resources	Minimise future financing needs	Harness additional sources of finance
Enhance the operational efficiency of water and sanitation service providers	Manage water demand	Ensure tariffs for water services reflect the costs of service provision
Encourage connections, where central assets are available	Strengthen water resource allocation	Consider new sources of finance
Develop plans that drive decisions	Encourage policy coherence across water policies and other policy domains (including nature-based solutions)	Leverage public and cohesion funds to crowd-in domestic commercial finance
Support plans with realistic financing strategies	Exploit innovation in line with adaptive capacities	
Strengthen capacity to use available funds		
Build capacity for economic regulation		

Recommendations need to be tailored to national and local contexts. They can be informed by good international practices. Some crosscutting messages deserve particular attention.

Planning has an essential role to play to ensure efficient allocation of finance. While plans are abundant in EU member states, they vary in their capacity to drive investment decisions. Investment planning should factor in demographic trends, including depopulation of rural areas and smaller towns. They need to reflect robust projections on climate change. Effective plans must be consistent with initiatives in other sectors (e.g. urban planning and land use; environment, agriculture, energy and transport). Plans are best accompanied by realistic financing strategies, which specify where finance will come from.

In selected areas, such as mountainous and isolated territories, cost-effective decentralised wastewater collection and treatment can be considered. Compliance monitoring and enforcement will be crucial to ensure environmental protection (i.e. to prevent groundwater contamination from leaking septic tanks, and inappropriate wastewater disposal without treatment to rivers). IAS should be considered in the context of national strategies, with mechanisms to ensure reliable performance of services. This is likely to increase the costs of IAS, in places making the connection to existing pipes competitive.

Independent economic regulation (usually at national level) can support the transition towards sustainable financing strategies for water supply and wastewater collection and treatment. Key features of well-defined independent regulation are to separate functions and powers of policy from operations, and to incentivise greater performance and accountability from local authorities, operators of water services and water users. Such oversight can strengthen the transition to cost recovery through tariffs, and stimulate improved performance of service providers (be they public or private).

Most countries will need to consider new sources of finance. Private finance (commercial debt or equity) is available in all EU member states. So far, it has only marginally been mobilised to finance investments in water supply and sanitation. There is room for manoeuvre to attract commercial capital for creditworthy borrowers to finance water-related investments. This may require exploring how public budgets, including cohesion policy funds and risk-mitigation instruments (e.g. guarantees, credit enhancement instruments) can be used strategically to improve the risk-return profile of investments that can attract commercial finance. Lessons from innovative arrangements in Europe to combine or blend public and private sources of finance could inspire further developments.

A role for the European Commission

On of all these and related issues, the European Commission has a role to play to foster and accelerate the transition towards robust strategies to ensure compliance with the DWD, UWWTD and FD, now and in the future. While Cohesion Policy has been a major driver for compliance across member states, other tools and mechanisms need to be considered and developed to adjust support to the distinctive needs and capacities of heterogeneous member states, particularly in the context of the decline of Cohesion Funding for water-related expenditures.

Most importantly, the European Commission needs to continue efforts to enforce compliance monitoring and reporting. This creates a momentum for data collection, strategic planning, and enhanced accountability of member states (to the European Commission and citizens).

Considering its importance for the cost-effectiveness of policy responses, the performance and efficiency of water supply and sanitation services deserve particular attention. This line of work potentially combines guidance on independent economic regulation (for tariff setting, benchmarking the performance of water utilities), agglomeration of small entities and support to define and implement robust national strategies for IAS. This can be achieved through a combination of peer learning and some form of conditionality of support.

Similarly, national and local authorities across Europe would gain from enhanced capacity to design and implement investment plans and expenditure programmes that contribute to compliance with the EU acquis at least cost for the community. As stressed above, such plans and programmes must consider long-term issues (including climate change) and coherence across policy areas. The new enabling conditions to access EU funds go in this direction, when they encourage robust investment plans combined with financial strategies. More guidance may be required to characterise appropriate plans and strategies.

In addition, member states would continue benefitting from practical policy guidance on pending and emerging issues, where member states are still looking for appropriate policy, technical and financial responses (e.g. CECs, combined sewer overflows, sludge). Any sort of guidance on these topics should cover issues related to costs and financing.

Other parts of the European Commission can also contribute. For instance, the development of domestic financial institutions can crowd-in private finance for water-related investments. Such institutions could usefully be encouraged as financial partners in the disbursement of EU Funds, whenever feasible. Support can be initiated by the European Commission, for instance in the context of the action plan on sustainable finance.

These options need to be refined and adjusted to reflect the priorities and means of action of the new European Commission. They provide a fertile ground to rejuvenate constructive and fruitful policy guidance and peer learning with and across member states, on water and related issues.

Background and process

The European Commission and the OECD endeavoured to assess the capacity of member states to cover the investment and financing needs they face now and by 2050 related to water supply, sanitation and flood protection. The assessment will inform strategic thinking about countries' investment planning and financing strategies as well as about the role that the European Commission can play to support its member states.

When assessing investment needs for water supply and sanitation and flood protection, three specific directives merit particular attention: the Urban Waste Water Treatment Directive (91/271/EEC); the Drinking Water Directive (98/83/EC); and the Floods Directive (2007/60/EC). Beyond these technical directives, the Water Framework Directive (2000/60/EC) provides framework legislation to facilitate the co-ordination of objectives and means of implementation for water-related policies and regulations.

Despite an overall relatively high level of compliance with the Drinking Water and Urban Waste Water Treatment Directives, lack of funding is mentioned by several countries as an – or the main – obstacle to achieve full compliance. Further, member states will have to face new challenges, which will call for additional investments in the water sector. Such challenges include emerging pollutants – which can require more stringent standards for drinking water and treated wastewater - and climate change.

Against this backdrop, the OECD collaborated with the European Commission to review future investment needs and financing capacities in the twenty eight EU member states, in order to establish a comparable overview across countries and identify those facing the most severe situations.

The review was organised along two phases. The first phase of the project assessed the EU 28 member states' investment needs by 2050 for water supply, sanitation and flood protection, as well as their capacity to finance these needs. In the second phase of the project, investment needs as well as financing strategies and options were looked at in more depth in nine EU member states likely to face the greatest challenges in financing their future investment needs. Recommendations from this second phase, while tailored to countries facing most severe financing challenges, are likely to be relevant for all EU member states and beyond.

Likewise, lessons on challenges and sustainable financing strategies from both phases of the work are relevant far beyond the EU - and indeed OECD - member states. These will be shared and discussed in the context of OECD work on financing water, including in meetings of the OECD Working Party on Biodiversity, Water and Ecosystems, and the Roundtable on Financing Water[1]. The Roundtable is designed as a collaborative platform to identify and promote concrete options to improve the mobilisation of capital towards investment in sustainable, resilient water infrastructure.

This report captures the main messages from the project. It consists of five parts:

- Part I frames the issue and sketches the drivers considered for the projections of investment needs.
- Part II characterises the state of play and member states' current level of effort to finance water supply, sanitation and flood protection.
- Part III presents the projections for future investment needs across member states.

- Part IV discusses the financing capacities of member states and signals particular challenges selected countries are likely to face to finance projected investment needs.
- Part V explores options that can help countries to minimise their financing needs for water supply, sanitation and flood protection as well as possibilities to harness additional sources of finance for flood protection. They reflect discussions that took place during seminars in nine countries facing the most severe financing challenges.

Notes

[1] http://www.oecd.org/water/roundtable-on-financing-water.htm

1 Framing the challenge

The chapter explains how compliance with the EU Directives on water contributes to sustainable growth. It identifies some of the main drivers of investment needs for water supply, wastewater collection and treatment and flood protection in Europe. It zooms on contaminants of emerging concern as an example of drivers which will affect investment needs but which is difficult to quantify.

This part reiterates the benefits of investing in water and protecting populations and ecosystems from the risks of too much, too little, polluted water and of a lack of access to safe drinking water and sanitation services. Against this backdrop, it characterises the ambition of this review, describes its scope and sketches the method used. This Part also presents and discusses the drivers, which have been identified and used to project expenditure needs.

1.1. The benefits of investing in water

The OECD (2018) argues that water-related risks increasingly affect stability and economic growth, public finances, poor and vulnerable social groups as well as the environment (see Box 1.1 for a definition of water-related risks and water security). In Europe, the BLUE2 report quantifies the value added and jobs in the water sector and in water-dependent sectors: the EU's water-dependent sectors generate EUR 3.4 trillion or 26% of the EU's annual Gross Value Added (Spit et al., 2018). Notably, the specific water-dependent sectors and the moderately dependent sectors contribute the most to the EU economy, each accounting for around 10% of total EU Gross Value Added. Moreover, EU's water-dependent sectors employ around 44 million people, i.e. 24.2% of the total employment, and include 16.3 million enterprises (for more information, see Spit et al., 2018).

Water-related risks demand urgent and concerted action. As populations, cities and economies grow and the climate changes, greater pressure is being placed on water resources, increasing the exposure of people and assets to water risks and the frequency and severity of extreme climatic events. If not properly addressed, rising water stress and increasing supply variability, flooding, inadequate access to safe drinking water and sanitation, and higher levels of water pollution will continue to act as a drag on economic growth.

In its analyses to restore EU competitiveness, the EIB claims that droughts have caused EUR 86 billion damages over the last 30 years. The costs of floods is even higher and amounts to EUR 150 billion in 2002-2013, the largest source of GDP losses from natural disasters in Europe (EIB, 2016). Jongman et al. (2014) projects that annual damages could multiply by four between 2014 and 2050 (from EUR 5.5 billion to EUR 23 billion).

Water security affects industries and their global value chains. It also affects energy security, as 44% of water abstraction in Europe is destined to energy production (hydropower, coal, or nuclear power). It obviously affects agriculture as well, which can use up to 80% of water in Southern regions (EIB, 2016).

> **Box 1.1. Water security, defined**
>
> The OECD defines water security as achieving and maintaining acceptable levels for four water risks:
>
> - Too little water (including droughts): Lack of sufficient water to meet demand for beneficial uses (households, agriculture, manufacturing, electricity and the environment);
> - Too much water (including floods): Overflow of the normal confines of a water system (natural or built), or the destructive accumulation of water over areas that are not normally submerged;
> - Too polluted water: Lack of water of suitable quality for a particular purpose or use; and
> - Degradation of freshwater ecosystems: Undermining the resilience of freshwater ecosystems by exceeding the coping capacity of surface and groundwater bodies and their interactions.
>
> These risks to water security can also increase the risk of (and be affected by) inadequate access to safe water supply and sanitation.
>
> The water risks are inter-related. For example, floods and droughts both affect water quality, the provision of safe drinking water, and contribute to degradation of freshwater ecosystems. Polluted water resources, without treatment, are effectively excluded from human consumption and utilisation by industry and agricultural sectors, thereby increasing the risk of water scarcity. Climate change is exacerbating existing water risks, due to altered precipitation and flow regimes, more frequent and severe extreme weather events, altered thermal regimes, and sea level rise. Moreover, the inherent uncertainty in climate change projections makes it more challenging to assess how these risks will evolve in the future.
>
> Investment in water security can help to safeguard growth against growing water risks. Decision makers will need to innovate and adapt, without being limited to the solutions adopted in the past.
>
> The OECD risk-based approach of "Know the risks", "Target the risks" and "Manage the risks" can assist in prioritising and targeting water risks, determining the acceptable level of risk, and designing policy responses that are proportional to the magnitude of the risk.
>
> Source: Adapted from OECD (2013a), Water Security for Better Lives, OECD Studies on Water, OECD.

Water security affects all countries in Europe, with the greatest threats of water-related risks falling mainly on countries in transition to advanced economies (Box 1.2). Globally, less developed countries face unreliable water supplies, and hence require greater investment to achieve water security. Developed countries – despite being relatively water secure - must continuously adapt and invest to maintain water security in the face of climate change, deteriorating infrastructure, economic development, demographic change, and rising environmental quality expectations.

The benefits of investment in water security are manifold. Investment in water security protects society and sectors from specific water risks, and can have a profound positive effect on economic growth, inclusiveness, and the structure of economies. For example, enhancing water security can reduce the price - and the price volatility - of staple food crops, a key priority in the global economy.

Across Europe, regulation provides incentives and guidance to drive water-related investment that contributes to sustainable growth. Three specific directives merit particular attention: the Urban Waste Water Treatment Directive (91/271/EEC); the Drinking Water Directive (98/83/EC); and the Floods Directive (2007/60/EC). Other Directives provide additional incentives, for instance to address diffuse pollution from nitrates. Ultimately, the Water Framework Directive (2000/60/EC) sets the overall level of ambition and facilitates the co-ordination of objectives and means of implementation for water-related policies and regulations.

Urging EU member states to invest in water security and supporting those most in need are essential contributions to sustainable and inclusive growth across the region. This assessment endeavours to understand the issues and opportunities EU member states face when it comes to investing in water security. It provides robust comparisons across countries, to characterise country-specific situations and challenges and support future discussions on options to overcome these challenges.

> **Box 1.2. Relative economic impacts of water insecurity**
>
> The *Global Dialogue on Water Security and Sustainable Growth*, a joint initiative by the OECD and the Global Water Partnership, examines the causal link between water management and economic growth.
>
> Different parts of the world are subject to different water risks, and many countries suffer from all water risks. Some countries are more vulnerable to water risks than others. A country's hydrology, the structure of its economy, and its overall level of wealth (and associated level of water infrastructure and institutional capacity), are all key determinants of its vulnerability to water risks.
>
> The risk of water scarcity is concentrated in locations with highly variable rainfall and over-exploitation of relatively scarce resources. Given that the dominant use of water is for agricultural irrigation (global average is 70%), the economic consequences of droughts and water scarcity are most pronounced in agriculture-dependent economies.
>
> The economic risks from flooding are increasing in all locations worldwide, due to increasing economic vulnerability, but are greatest in North America, Europe and Asia.
>
> The greatest economic losses are from inadequate water supply and sanitation, and associated loss of life, health costs, lost time, and other opportunity costs. The losses are greatest in Sub-Saharan Africa.
>
> China and India suffer the greatest total economic burden, and number of people at risk, of water insecurity, and are subject to risks of water scarcity, floods, and inadequate water supply and sanitation.
>
> Different parts of the world are subject to different water risks, and many countries suffer from all water risks. Some countries are more vulnerable to water risks than others. A country's hydrology, the structure of its economy, and its overall level of wealth (and associated level of water infrastructure and institutional capacity), are all key determinants of its vulnerability to water risks.
>
> The risk of water scarcity is concentrated in locations with highly variable rainfall and over-exploitation of relatively scarce resources. Given that the dominant use of water is for agricultural irrigation (global average is 70%), the economic consequences of droughts and water scarcity are most pronounced in agriculture-dependent economies.
>
> The economic risks from flooding are increasing in all locations worldwide, due to increasing economic vulnerability, but are greatest in North America, Europe and Asia.
>
> The greatest economic losses are from inadequate water supply and sanitation, and associated loss of life, health costs, lost time, and other opportunity costs. The losses are greatest in Sub-Saharan Africa.
>
> China and India suffer the greatest total economic burden, and number of people at risk, of water insecurity, and are subject to risks of water scarcity, floods, and inadequate water supply and sanitation.

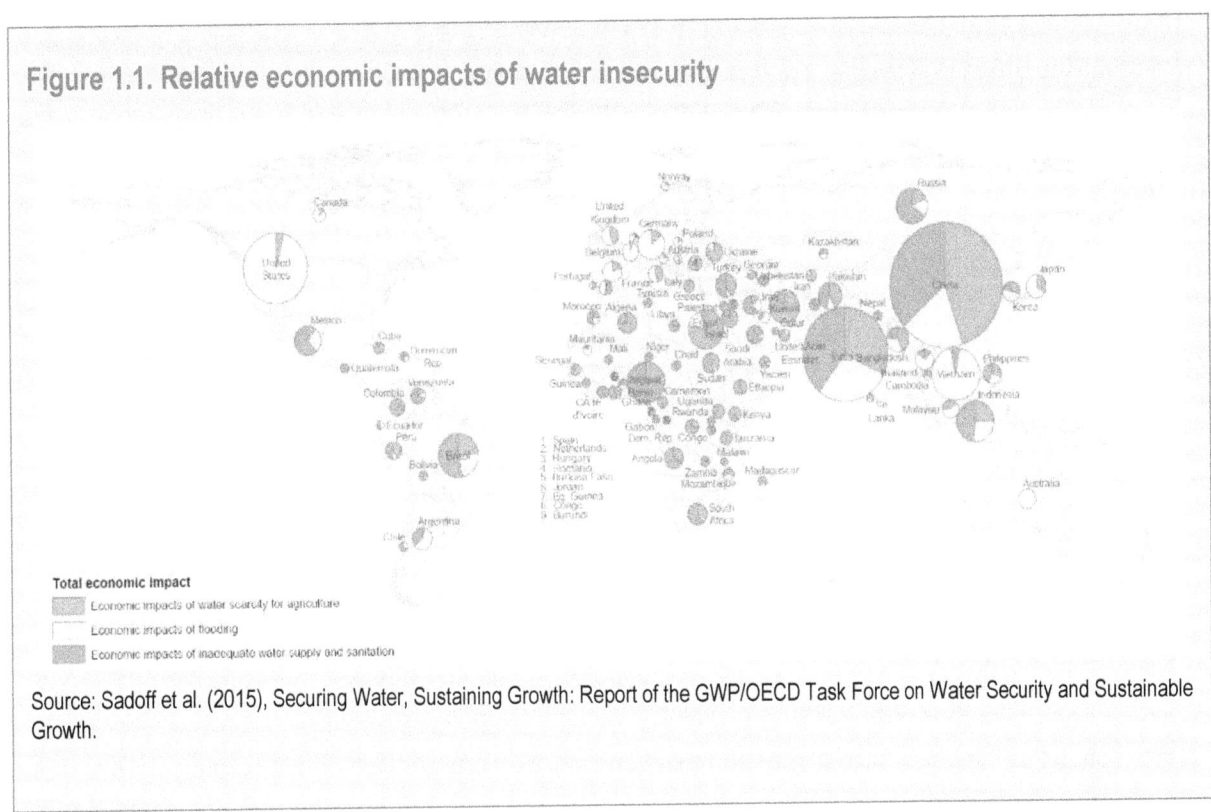

Figure 1.1. Relative economic impacts of water insecurity

Source: Sadoff et al. (2015), Securing Water, Sustaining Growth: Report of the GWP/OECD Task Force on Water Security and Sustainable Growth.

1.2. Ambition and scope of the project

The overall ambition of the project is to assess the capacity of the 28 EU member states to cover the investment and financing needs they face now and by 2050 related to water supply, sanitation and flood protection. Investment needs to comply with the Water Framework Directive cannot be projected using the same method and will be addressed separately. The study accounted for the outermost regions of the EU in estimation of the investment needs to the extent that these regions are captured by national data reported to Eurostat. The analysis did not consider the specific situations of these outermost regions in terms of financing options for the future.

Projections of future investment needs derive from a baseline of current expenditures (based on best-available and comparable data) and the influence of several drivers of investment needs. Three scenarios are considered. The drivers of investment are discussed in the following section.

- *Business as usual for water supply and sanitation services and flood protection*. This scenario projects the same level of effort, with no new policies. Projections are driven by urban population growth (see below the discussion on drivers).
- *For water supply: projections to achieve compliance, efficiency and access*. Most EU member states already comply with, or are close to complying with, the Drinking Water Directive (DWD). The revised DWD will trigger additional investment needs, which is reflected in the projections, building on an assessment made by the European Commission. It is anticipated that, even when member states comply with the revised DWD, countries will need to invest in water efficiency and minimise non-revenue water (including leakage). In addition, countries will have to ensure that vulnerable groups have access to safe water. The additional costs of providing access to these groups has been assessed by the European Commission and reflected in this scenario.

- *For sanitation: projections to achieve compliance.* Several EU member states do not fully comply with the Urban Wastewater Treatment Directive (UWWTD). The extent of compliance varies across EU member states and has been considered the main driver for additional investment in this domain.

The projections need to be qualified in several ways. First, the business as usual scenario and the projections reflect the current level of effort. They do not consider the potential delay or backlog of investment and the state of existing infrastructures. This is an important caveat for water supply and sanitation services, typically, where current levels of efforts in many countries may not allow for proper maintenance and renewal of existing assets. This may explain why country specific assessments (when they consider the state of the asset and the investment backlog) may differ from the projections made in this report.

Second, the current level of efforts in flood protection was not monetised. Only a few countries monitor financial flows for flood protection, usually the ones who can be expected to spend the most (Austria, the Netherlands). It was not possible to extrapolate on the basis of available data. Therefore, projections on investment needs for flood protection are based on changes in the exposure to flood risks.

Third, emerging challenges, which could not be monetised, are discussed qualitatively. These include climate change and contaminants of emerging concern (e.g. focused primarily on pharmaceuticals for the purpose of this analysis). A rough estimate of investment needs to address contaminants of emerging concern is presented at an aggregate level, using costs measured in Switzerland.

Fourth, options to minimise financing needs exist and will be considered by most countries. This is the case, notably, of i) distributed systems and a range of innovative ways to build, manage and finance water supply and sanitation systems; and ii) nature-based solutions for sanitation and for flood protection. How these options will materialise and affect investment needs in each member state remains highly uncertain. Therefore, such options are discussed in the report, but not reflected in the monetised projections.

Finally, compliance with the Water Framework Directive is not properly captured in the projections: it requires a range of very diverse measures. Therefore, it is difficult to track expenditures that contribute to compliance with the WFD. Moreover, cross-country comparisons of expenditures and costs are unlikely to provide valuable information. The European Commission is considering additional research in 2020-21 to assess how countries implement the economic and financing dimension of the WFD. This report discusses selected issues related to the WFD separately.

Another challenge is not covered here: securing sufficient water resources to meet demand. Supply augmentation, abstraction and production of bulk water are not covered by this project. While this issue gains prominence across EU member states and the costs can be significant, there are too many uncertainties on how countries will address it to substantiate any robust discussion of costs and financial requirements at regional level. For instance, in the UK, the National Infrastructure Commission favours a twin-track approach that combines supply augmentation – via additional reservoirs and reuse and potentially a national water network - with demand management – via leak reduction and systematic roll-out of smart meters (NIC, 2018). Total costs will heavily depend on how these different options are balanced and combined.

The method and data used to support the baseline and the projections are synthesised in Annex B. They are described in more detail in a separate methodological note.

1.3. Drivers of investment needs

Drivers of investment needs in water security are wide-ranging and context-dependent (see Box 1.3 below). What is considered to be an acceptable level of water risk in a given country strongly shapes investment needs. This can shift over time. Generally, as economies develop, the tolerance for water-related risks declines (OECD, 2013a).

> **Box 1.3. Experts' view on future drivers of water infrastructure investment needs**
>
> The Report of the OECD-World Water Council *High-Level Panel on Financing Infrastructure for a Water-Secure World* (Winpenny, 2015) compiles the best available knowledge about future investment and water-related expenditures. The report acknowledges that projections in this area are particularly difficult.
>
> A Delphi survey shed some light on the main drivers for future water infrastructure needs:
> - Social perception of - and responses to - water-related risks (in particular droughts, floods, pollution)
> - Awareness of the value of ecosystems and biodiversity
> - Innovation in water services and infrastructure; and
> - How changes in climate affect water availability and demand.
>
> In a European context, Cambridge Econometrics (2017) reports that main drivers for investment in water supply and sanitation considered by stakeholders are compliance with EU policy, maintenance of sustainable services and higher efficiency in service delivery.
>
> In that context, projections of future investment needs depend on a range of definitions and choices, and these are difficult to compare.
>
> Source: Winpenny J. (2015), Water: fit to finance? Catalyzing national growth through investment in water security, Report of the High-Level Panel on Financing Infrastructure for a Water-Secure World, World Water Council and OECD. Cambridge Econometrics (2017), Bridging the water investment gap, a report to the European Commission DG Environment.

This assessment selects a range of drivers, which are briefly described in this section. They can be identified based on the type of water-related risk.

- Water supply
  - Urbanisation (and the number of additional people to be connected to water supply systems)
  - Compliance with the Drinking Water Directive
  - The number of people who do not have access to water Additional investment to approximate the best performance in terms of water networks efficiency (minimising non-revenue water or resource losses).
- Sanitation
  - Urbanisation, i.e. the number of additional people to be connected to sanitation systems
  - Compliance with the Urban Wastewater Treatment Directive
- Flood protection
  - The value of assets at risk of flooding.

1.1.1. Drivers selected for projections related to water supply and sanitation

In an earlier OECD study (OECD, 2006), Ashley and Cashman projected water-related investment needs as a share of GDP, acknowledging economic growth as a major driver for water-related investment needs. This report builds on this earlier OECD work and expands the range of drivers beyond economic growth to reflect the situation in EU member states. The following text justifies the main drivers.

Demographics and urban population growth

Demographics is known to be a major driver for growth. It is also a major driver for investment in water supply and sanitation, as it dictates the number of people to be connected to services. In the European Union, where a vast majority of people live in urban areas, urbanisation continues to drive investment needs in water supply and sanitation. It also drives the value of assets at risk of flooding.

Currently, on average across the twenty eight member states, 96% of EU citizens are connected to potable water supplies (only 57% in Romania) – the highest connection rates to date. The overall proportion of citizens connected to water supply services is expected to remain stable to 2050 (EC, 2017). This is due to lifestyle as well as living conditions and location (remoteness) in a number of member states, which are not conducive to connections to water supply networks at an affordable cost. A relatively low connection rate may either indicate that selected groups (in particular vulnerable ones) do not have access, or that connection may not be appropriate in places in remote locations (think of Sweden and secondary homes on islands).

Despite a stable proportion of the population maintaining access to water services, overall a greater number of people will gain connection in the future. This derives from the fact that the total population increases in most EU countries.

Figure 1.2 illustrates the high variability of the extent of urbanisation between selected EU member states. The extent of the impact of urban population growth on expenditure needs for water supply and sanitation may depend on the current capacity usage of already installed infrastructure. Some countries (e.g. Germany) enjoy excess capacity, which will allow to absorb urban population growth without (or with limited) extension of existing networks. In contrast, other countries or cities have reached full capacity and any growth in urban population will require additional construction of reservoirs, pipes and treatment facilities (e.g. Dublin, Ireland). In the case of Romania, while the share of urban population is expected to increase, total population is forecasted to decrease.

Figure 1.2. Share of population residing in urban areas (%)

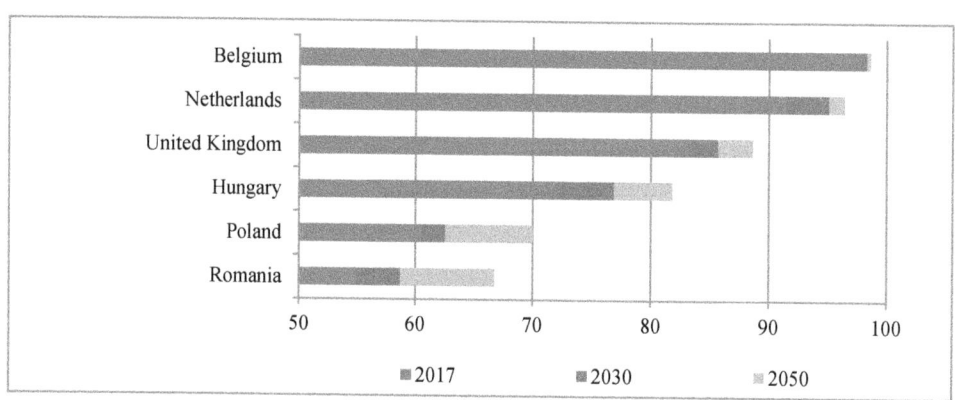

Source: United Nations Department of Economic and Social Affairs (consulted in 2018).

In addition, as the urban population increases, it is anticipated that the number of people who face potential health risks from water-related disease outbreaks in public waters will remain significant, though slightly decreasing from an estimated 22.7 million in 2015 to 20 million in 2050, equivalent to 4% of EU 28 population in 2050. The countries with the greatest proportion of population potentially at risk are: Bulgaria (12%), Romania (8%) and Belgium (9%); whereas the highest numbers of citizens potentially at risk are found in: Italy (3.4m), Spain (3.3m) and Germany (2.7m). Even the best performing countries still have substantial numbers of citizens potentially at risk of water-related disease outbreaks; e.g. the UK (circa 800,000 in 2015). These numbers are expected to decline in most EU member states by 2050 (EC, 2017).

Compliance with the Drinking Water and Urban Wastewater Treatment Directives

Despite an overall high level of compliance with the Drinking Water Directive, some countries may lack funding or face unsustainable financing strategies to achieve and maintain full compliance. This is particularly the case as the proposal for the revised Drinking Water Directive[1] considers more stringent standards and increased access to water for vulnerable groups.

Compliance with the UWWTD is high, but several countries are still lagging behind and more efforts are required to reach full compliance. Indeed, a number of countries project additional investment and expenditures to reach compliance in the coming few years[2].

The efficiency of water supply services

In addition to additional population to be served and new regulations, all member states face additional challenges related to operating, maintaining and upgrading existing assets and improving the efficiency of water networks. As new assets are built and existing infrastructure ages, the recurrent expenditures to operate and maintain them increase. The efficiency of asset operations and maintenance and effectiveness of recurring expenditures dictate the capacity of existing assets to deliver reliable service over time and the need for renewal in the future. The Figure 1.3 below provides a stylised illustration of the volume of investment needs and the share allocated among different types of expenditures, as the infrastructure develops. At the beginning of a cycle, countries invest in capital-intensive installation of new networks and equipment. Then, upfront capital requirements decrease while the costs of operating and maintaining installed capacities increase. This requires a shift to a different type of expenditure, often financed through distinct instruments: while upfront investment is traditionally financed through public finance, operation and maintenance (O&M) of existing assets are often financed by the revenues of tariffs that reflect the cost of service provision (Ireland is an exception). As infrastructures age, O&M cost increase to a point where investment in renewed assets is required. It remains to be seen how improved O&M and maintenance translate in terms of current and capital expenditures over time.

The Figure below illustrates this sequence, assuming that different countries (countries 1 to 7) kick-start the cycle at different points in time. This phasing reflects the fact that different countries join the European Union in different years and embark in the cycle of complying with the DWD and UWWTD at different times.

Figure 1.3. Stylised sequence of investments in water supply, sanitation and flood protection

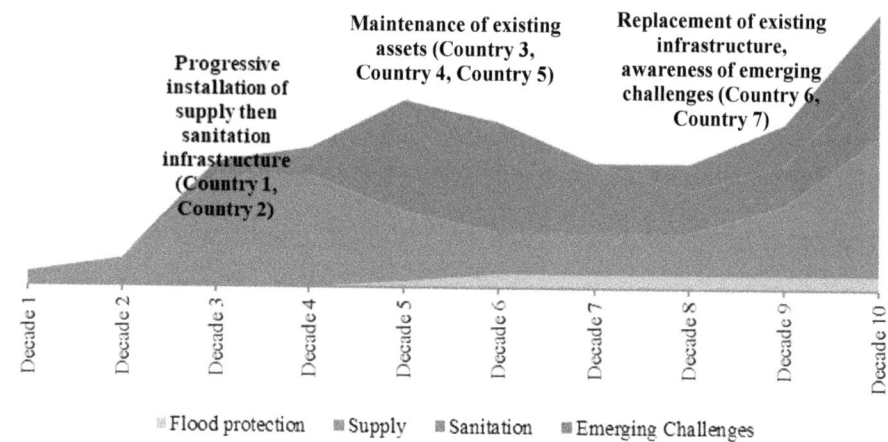

Source: Authors.

For both water supply and sanitation, efficiency gains will remain one of the biggest challenges in EU member states. Asset deterioration results in leakages and decline in water quality, affecting the health of human and ecosystems (both surface and groundwater) and increasing treatment costs downstream.

In its first report on the issue, the European Association of Water Regulators (WAREG, 2017) uses key performance indicators (KPI) to assess the efficiency of services in Europe. The report confirms that there is no single definition of efficiency and KPIs vary (see the definitions from IWA (2016), IBNet - https://www.ib-net.org/toolkit/ibnet-indicators/). Moreover, the report documents the range of operating environments across Europe. As a matter of illustration:

- Total volume of water sold per person per day ranges from 80 to 234 litres
- Non-revenue water ranges from 17 to 67% of net water supplied. Distribution losses reported by EurEau (2017) vary between less than 10% (in the Netherlands, Germany and Denmark) and more than 40% (in Malta and Ireland).
- The total number of breaks per km of pipes varies between 0.1 and 4.43 breaks/km/year
- The number of staff employed in utilities varies between 1.2 and 8.64 employees per thousand connections.

Such variations reflect geography and history of water supply and sanitation services. France, for instance, has almost 1 million kilometres of pipes on its vast and low-density territory; twice as much as Germany, the UK, or Italy (see EurEau, 2017). Variations also reflect the ability to operate, maintain and renew assets.

Water main leakage and breakage rates are not reduced at significant rates in member states. The situation is especially striking in Hungary, where water losses have increased by 5% between 2005 and 2013 and now reach 26%; Romania, where losses increased from less than 30% to almost 40%; and Bulgaria, where they remain stable at 60% (European Court of Auditors, 2017). The data for Portugal illustrate that the majority of losses results from the distribution network of smaller pipes (retail network) rather than from the transmission network of larger pipes (bulk water systems) (WAREG, 2017).

In addition to investment in renewed infrastructures to reduce leakage, effectively reducing non-revenue water will require investment in data management systems for customer billing, bill collection activities, metering of all uses (including construction and firefighting), and establishment of regular auditing procedures. This will result in operational and capital cost increases across multiple facets of the service provision.

Compliance with the Water Framework Directive

The overall objectives of the EU's water policy are set by the WFD, which aims at the non-deterioration and achievement of good status of all EU water bodies. The UWWTD and the DWD are basic (i.e. compulsory) measures under the WFD. The FD is complementary to the WFD and sometimes leads to trade-offs.

Compliance with the UWWTD certainly contributes to good status, as a series of contaminants are collected and treated before treated wastewater returns to the environment. However, other contaminants may remain in treated wastewater, and additional wastewater treatment is a supplementary measure under the WFD. Moreover, good ecological status also refers to other dimensions; for instance, in the case of surface water, they include the hydromorphology of river bodies.

Compliance with the Water Framework Directive will depend on mitigation of pressures, such as the reduction of nitrates and diffuse pollution from urban or agriculture runoff. It will also depend on the re-naturalisation of rivers and lakes, and clean-up of historic contamination. The cost of such measures to comply with the Water Framework Directive is difficult to estimate, especially in a way that allows for cross-country comparisons[3]. Therefore, the additional cost of complying with the Water Framework Directive could only be discussed qualitatively in the context of this assessment. This discussion is captured in Part III below.

Box 1.4. Water mains and sewerage infrastructure renewal in England and Wales

Current replacement rates for sewers in England and Wales are significantly below what is typical for sewers. The rate of expenditure for water mains and sewer renewal rates is £962m (EUR 1,100 million) per annum for the current 5-year investment period, which is estimated to lead to pipe networks beginning to fail more often, affecting water customers and the environment.

A study by UK Water Industry Research (2017) predicts that by 2050:

- the number of water main bursts will increase by 20%
- the number of interruptions to water supplies will increase by 25%
- leakage will increase by 40% unless other leakage control measures are significantly increased
- sewer blockages and collapses, and the resulting flooding and pollution, will increase by 6%.

If the situation continues to deteriorate, it will become more expensive to recover from. For example, if there is no increase in expenditure for the next 10 years, then an additional £11 billion (EUR 12.5 billion) will be needed to return the networks to the condition needed to meet current service standards.

Table 1.1 shows the required increases in renewal rates and associated relative expenditure.

Table 1.1. Rates of expenditure for "best estimate" scenarios

	Current (2015-2020)	Short term (2020-30)	Long term (2030-70)
WATER MAINS			
% annual renewal	0.6	1.2	1.3
Cost per unit length (index)	100	100	152
Expenditure (index)	100	200	330
SEWERS			
% annual renewal	0.2	0.8	1.2
Cost per unit length (index)	100	100	113
Expenditure (index)	100	400	680

The study also examined potential new technological developments that will reduce expenditure needs by reducing the rate at which both new and existing pipes fail, reducing the cost of new pipes (or pipe rehabilitation), or reducing the consequence of pipe failures. The study concluded that incremental improvements are expected in the area of new materials development, but their impact is likely to be small regardless of the year they are adopted. The highest expected impact would come from further development of the ability to predict performance via continuous monitoring.

Source: UKWIR, 2017.

1.1.2. Drivers selected to project expenditures for flood protection

In many European countries actual flood risk is expected to increase in the future due to climate change and socio-economic developments. On the one hand, flood probabilities are expected to increase due to climate change induced impacts on river discharges, sea level rise and extreme weather events. On the other hand, flood damages are also expected to increase due to economic and population growth. Consequently, investments in measures to flood risk will be required to maintain current flood protection levels in the future.

For the purposes of this report, flood protection investment need is defined as the financial resources that are required to maintain actual (existing) flood risk (and the corresponding flood protection standards of flood defenses) at the same level until 2050. Upgrading of flood protection standards through new flood policies is not included in the analysis.

A risk-based approach towards flood risk management is adopted, in which risk is defined as the flood consequences (damages, victims) multiplied by the probability of flooding. For projection of river flood risk investment needs, it is assumed that future investment needs will follow the same pace as changes in river flood risk due to climate change and socio-economic developments.

Projecting expenditure needs to protect against riverine floods

As a first step to estimate projected river flood risk investment needs to 2030, the expected change in future flood risk is calculated as the difference between current river flood risk and river flood risk in 2030 due to climate change and socio-economic developments.

A country's level of flood risk is determined by existing flood protection standards, the corresponding expected economic damage (direct and indirect), and the corresponding expected number of victims (injuries and casualties). Therefore, changes in flood risk to 2030 are represented by three indicators:

1. Annual expected urban damage (indicator for the value of assets at risk - this represents the vulnerability to direct economic damage)
2. Value of exposed GDP (indicator for economic activity at risk - this represents the vulnerability to indirect economic damages)
3. Size of expected exposed population.

The scenario applied to study changes in flood risk indicators is a combination of severe climate change and continued current socio-economic development trends[4].

Projecting expenditure needs to protect against coastal floods

Very limited information is available on changes in vulnerability factors that compose coastal flood risk (exposed population in coastal floodplains, exposed GDP in coastal floodplains, exposed urban assets in coastal floodplains) at the EU-28 level. The information that is available is dispersed across different studies, using different assumptions and methodologies, scenarios and time horizons. Only a few studies are available that link projections of coastal flood risk to the vulnerability of coastal areas in terms of economic damages and victims.

In that context, coastal flood risk investment needs to 2030 were qualitatively projected, based on data for three indicators, documented by distinct papers:

1. Change of population density in areas vulnerable to coastal flooding (Brown et al., 2011): the percentage increase of built-up in areas vulnerable to coastal flooding.
2. Number of people exposed to flooding (Hinkel et al., 2010): expected number of people subject to flooding to 2050.
3. Damage costs in the case of a flood event (Hinkel et al., 2010): the annual costs of economic damage caused by the sum of coastal flooding, dryland loss, wetland loss, salinity intrusion and migration.

1.4. Emerging challenges

The section explores two challenges member states face – climate change and contaminants of emerging concern in water bodies - which will affect the costs of supplying water and sanitation services and of protecting against flood risks. Experience with such challenges is still limited and the options to address them vary in terms of costs. Nevertheless, the section signals that investment needs to overcome these challenges can be high, depending on the severity of the challenge, the ambition of the response, and the policies and technologies implemented to respond.

1.1.3. Climate change

Climate change is projected to increase investment needs relating to water. The impacts of climate change on exposure to flood risks are factored in projections of assets, GDP and population at risks of flooding.

The impacts of climate change on expenditure needs for water supply and sanitation are partially captured in the Business as Usual scenario, assuming countries already factor in some level of adaptation to climate change. The need to further adapt the level of service and the infrastructures to future uncertainty and variability of water demand and availability, driven by climate change, is only addressed qualitatively. It is discussed below.

In some regions, increased investment will be required to address less favourable hydrological conditions – declining rainfall and snowpack, increasing variability, and more floods and droughts. Climate change is also likely to impact water quality. For example sea-level rise is projected to extend areas of estuaries and increase salt-water intrusion of freshwater aquifers, resulting in a decrease of freshwater availability, and toxic algal blooms and the growth and survival of pathogens are projected to increase with increases in water temperature, posing greater risk to drinking water quality (OECD, 2017).

Even where conditions become more favourable, there may be transition costs in moving to water management systems that are fit for the new climate regime. See Box 1.6 for a characterisation of efforts to adapt water management to climate change in OECD countries.

In addition, the unprecedented rate of change and potential novel changes outside of historical experience introduce a greater degree of uncertainty beyond what water managers have traditionally had to cope with. This increases the costs of water management, as systems have to be robust to a broader range of potential hydrological conditions.

The EU project ECONADAPT (2015) states that the costs of retrofitting wastewater and stormwater infrastructure to cope with higher water flows under climate change can be high. Hughes et al (2010) estimated the costs of climate change adaptation for OECD countries by region: overall the adaptation costs as a proportion of baseline expenditure range from 0.8 – 3.6% for Western Europe, and 6% - 13% for Eastern Europe for two adaptation scenarios. However, parts of the water system are likely to be open to cost savings in Western Europe, as the costs of new or replacement of existing assets is not very sensitive to changes in the flow volumes conveyed. Hughes et al conclude that upfront investment in adaptation for water and wastewater can generate net positive benefits. Urban drainage systems are one area where adaptation investments can bring net positive benefits over time, alongside urban infrastructure.

The management of combined sewer overflows (CSO) due to heavy rains is a good illustration of the magnitude of the challenge and of the range of options available (see Box 4.3 for a review of options). Countries differ in their risk of exposure to heavy rains. Milieu (2016) clusters member states according to exposure to heavy rains:

> "Member States that are particularly at risk for the consequences of heavy rain are: Belgium, Croatia, Italy, Luxembourg, the Netherlands, Portugal, Romania and Slovenia. The list includes several Mediterranean countries, at risk for heavy rainfall, which may be intense, of short duration, following a dry period and potentially leading to flash floods and storm water overflows. Also in mountainous area, higher number heavy rain events can be expected and is expected to lead, where sewer collection systems are present, to storm water overflows. Based on the observed trends, northwest Europe (Ireland, Finland, Sweden, Estonia, Lithuania, and Latvia) has the lowest risk for heavy rainfall (though not the United Kingdom). "

Table 1.2. Projected impacts of climate change on water across Europe

European sub-region	Impacts of climate change on water
Northern Europe	Temperature rise much larger than global average
	Decrease in snow, lake and river ice cover
	Increase in river flows
	Northward movement of species
	Increase in crop yields
	Decrease in energy demand for heating
	Increase in hydropower potential
	Increasing damage risk from winter storms
	Increase in summer tourism
Mediterranean region	Temperature rise larger than European average
	Decrease in annual precipitation
	Decrease in annual river flow
	Increasing risk of biodiversity loss
	Increasing risk of desertification
	Increasing water demand for agriculture
	Decrease in crop yields
	Increasing risk of forest fire
	Increase in mortality from heat waves
	Expansion of habitats for southern disease vectors
	Decrease in hydropower potential
	Decrease in summer tourism and potential increase in other seasons
North-western Europe	Increase in winter precipitation
	Increase in river flow
	Northward movement of species
	Decrease in energy demand for heating
	Increasing risk of river and coastal flooding
Central and eastern Europe	Increase in warm temperature extremes
	Decrease in summer precipitation
	Increase in water temperature
	Increasing risk of forest fire
	Decrease in economic value of forests

Note: Specific impacts for mountainous regions and coastal and regional seas are not shown.
Source: EEA (2017).

Box 1.5. Managing combined sewer overflows

There are 650,000 combined sewers across the EU member states, according to EurEau (2016), which discharge untreated wastewater, including priority hazardous substances and other substances into the environment. The regulation of these should ensure that appropriate storage and flow constraints are put in place to at least moderate the potential impacts at proportionate costs.

One option is to separate existing combined sewers into a sanitary network and a stormwater network. The cost to separate all of the combined sewers in the USA is estimated at some US$40.8 billion (EUR 37 billion) at 2012 prices (USEPA, 2016), illustrating the scale of the investment. Still, separate sewer and storm water connections will still cause some level of pollution depending on the level of wastewater/storm water treatment. Therefore, separation should come with downstream treatment using wetlands, ponds, filtration or other suitable systems (in particular to control substances washed from pavements by rainwater). Accordingly, the Netherlands and Germany only promote separation programmes where suitable treatment for rainwater is provided.

Traditional ways of managing CSOs include increasing capacity for storm water storage (including underground storage chambers) to reduce the frequency and amount of overflows. But they are costly at some EUR 1000 or more per cubic metre of storage provided. This approach is still being used in many parts of the EU, for example, in the EUR 6 billion 'supersewer' for London now under construction in response to the requirements of the UWWTD. The unique financing model (OECD, 2017), has meant that the projected maximum additional cost to 'customers' is estimated at some EUR 22 - 28 per customer per annum by the mid-2020s (in 2015 prices).

An alternative approach is to prevent storm water from entering the sewer network by using green infrastructure. This is invariably cheaper, more adaptable and more resilient than the traditional subterranean storage approach. Certain green infrastructures are better able to handle pollutants. Unfortunately, in certain jurisdictions, the institutional and other arrangements are not conducive to adopting any option other than traditional grey infrastructure as is happening in London (Dolowitz et al., 2017).

Sources: as above; full citations provided in the references list.

> **Box 1.6. Efforts to adapt water management to climate change in OECD countries**
>
> Progress in adapting water systems to climate change has advanced rapidly in recent years and a significant number of efforts are currently on-going. Impacts on freshwater nearly always feature as a key priority on OECD national risk assessments or adaptation strategies.
>
> The majority of efforts to date have focused on documenting the risk by building the scientific evidence base and disseminating information, but much more can be done to better understand what an acceptable level of risk is for a given population under specific circumstances, and to manage water risks in a changing climate.
>
> In particular, only a handful of countries have begun to address the issue of financing adaptation for water systems. Of those countries that have started financing water systems adaptation, some are mainstreaming adaptation into existing budgetary mechanisms, while others are addressing adaptation via specific water programmes or projects or tapping international financing mechanisms. A few countries have allocated dedicated funding to climate change adaptation in general, which typically includes measures for water.
>
> Source: OECD (2013b), Water and Climate Change Adaptation. Policies to Navigate Unchartered Waters, OECD Publishing; OECD (2015), Climate Change Risks and Adaption: Linking Policy and Economics, OECD publishing.

1.1.4. Contaminants of emerging concern

Contaminants of emerging concern[5] (CECs) comprise a vast array of contaminants that have only recently appeared in water, or that are of recent concern because they have been detected at concentrations significantly higher than expected, or their risk to human and environmental health may not be fully understood. Examples include pharmaceuticals, industrial and household chemicals, personal care products, pesticides, manufactured nanomaterials, and their transformation products.

The section focuses on pharmaceutical residues, as several European and other OECD countries gain experience with policy responses. The section builds on OECD work on the issue. Future OECD work on CECs will focus on micro-plastics in 2019-2020.

Pharmaceutical residues in the environment are an emerging concern

Pharmaceuticals are essential for human and animal health. However, they are increasingly recognised as an environmental concern when their residues enter freshwater systems. Pharmaceuticals are present in the environment as a consequence of pharmaceutical production and formulation, patient use, use in food production and improper disposal. Untreated household wastewater and effluent from municipal wastewater treatment plants are the most dominant sources of pharmaceutical residues to freshwater bodies globally; however, emissions from manufacturing plants, hospitals, specialised health care facilities, and intensive agriculture and aquaculture can be important pollution hotspots locally.

The presence of pharmaceutical residues in the environment poses an increasing problem. The number and density of humans and livestock requiring healthcare is escalating. This problem is further exacerbated, particularly in high-income countries, by growing numbers of elderly people with chronic health problems, and an increase in disease related to climate change. With this comes an increase in the quantity and diversity of pharmaceuticals produced, consumed and subsequently excreted.

Active pharmaceutical ingredients are found in surface waters, groundwater, drinking water, soil, manure, biota, sediment, and the food chain. Because pharmaceuticals are intentionally designed to interact with living organisms at low doses, even low concentrations in the environment can have negative impacts on freshwater ecosystems.

For example, active substances in oral contraceptives have caused the feminisation of fish and amphibians; psychiatric drugs, such as Prozac, alters fish behaviour making them less risk-averse and vulnerable to predators; and the over-use and discharge of antibiotics to water bodies exacerbates the problem of antimicrobial resistance – declared by the World Health Organisation as an urgent, global health crisis that is projected to cause more deaths globally than cancer by 2050.

Advances in analytical methods and risk assessment provide opportunities to build a policy-relevant knowledge base

Currently, there are a number of uncertainties associated with the environmental risk assessment of pharmaceuticals due to lack of knowledge concerning their fate in the environment and impact on ecosystems and human health, and the effects of mixtures of pharmaceuticals. The cost of monitoring, limited data for policy development and an absence of a systematic approach to risk assessment were three barriers to taking action identified by governments in the 2017 OECD Questionnaire on Contaminants of Emerging Concern in Freshwaters. Most OECD countries have established watch-lists and voluntary monitoring programmes for certain pharmaceuticals in surface water, but the majority of active pharmaceuticals ingredients, metabolites and transformation products remain unmonitored.

Advances in monitoring technologies can help close the knowledge gap and support policy responses. Real-time in-situ monitoring, passive sampling, biomonitoring, effects based monitoring, non-target screening, hotspots monitoring, surrogate data methods, early-warning systems and holistic modelling can help identify and anticipate sources of contamination. Country and international initiatives are crucial to improve the knowledge base and exchange of data, methodologies and technologies between countries and sectors.

Potential costs of addressing CECs and freshwater. Lessons from Switzerland

Current policy approaches to manage pharmaceutical residues in water are often reactive (i.e. measures are adopted only when risks are proven and routine monitoring technologies exist), substance-by-substance (i.e. environmental quality standards for individual substances) and resource intensive. Diffuse pollution, particularly from livestock and aquaculture, largely remains unmonitored and unregulated. Such approaches are not adequate for current and emerging challenges.

Switzerland is the first country to tackle the CECs challenge at the national level. The Swiss response to the challenge is described below. This is a systematic approach, which comes at a cost. Annex A summarises data collected in the literature on the possible costs of managing CECs in water streams.

> **Box 1.7. Addressing pharmaceutical residues in freshwater. The Swiss approach**
>
> Switzerland has committed to remove 80% of CECs from wastewater by 2040. The Swiss Waters Protection Act requires polluted wastewater produced by households, businesses or industry to be treated before being discharged into water bodies. In 2014, the Waters Protection Act was revised, to further improve wastewater treatment for the removal of CECs (including pharmaceuticals). The revised Act involved three policy instruments: i) a new technical wastewater treatment standard, and ii) a nationwide wastewater tax, and iii) public subsidies to fund technical upgrades of WWTPs. The technical standard requires selected WWTPs to remove 80% of CECs from raw sewage, measured on the basis of 12 indicator substances, by 2040.
>
> The standard applies only to WWTPs that meet one of the following three selection criteria:
>
> - Large WWTP servicing > 80,000 population equivalents (hereafter, p.e.);
> - Medium-size WWTP (24,000-80,000 p.e.) that discharge into small rivers with low dilution ratio; and/or
> - Medium-size WWTP (24,000-80,000 p.e) that discharge into water bodies used for drinking-water purposes.
>
> In total, approximately 120 out of 700 WWTPs met one of the above three criteria for upgrade. It is projected that this will result in a 50% overall load reduction of CECs in surface water. In addition, several WWTPs will be closed and wastewater transferred to larger facilities where the treatment is considered to be more cost effective.
>
> Pilot- and full-scale facilities assessed the effectiveness of various advanced wastewater treatment technologies, including ozonation, powered activated carbon, granular activated carbon, high-pressure membranes and advanced oxidation processes. Ozonation and powdered activated carbon showed the best applicability for Switzerland with the two techniques combined capable of removing 80% of detected CECs in wastewater.
>
> The total investment cost to equip 100 WWTPs with advanced treatment technology was estimated to be CHF 1.2 billion (EUR 1.1 billion). Operation and maintenance costs were estimated to be an additional CHF 130 million (EUR 119.6 million) per year, equivalent to 6% of the total current cost of wastewater treatment in Switzerland annually. The majority of the costs (75%) are financed by a new nationwide wastewater tax of CHF 9 (EUR 8.3) per person per year, which is earmarked in a federal fund to upgrade WWTPs. The remaining 25% of costs are covered by the municipalities. As WWTPs are upgraded and become operational, the municipalities are exempted from the tax.
>
> Despite having higher estimated costs than preventative source-directed measures, the end-of-pipe approach was selected because it was more predictable, measurable and feasible, and received support from industry, business, farmers, the research community and international actors. Furthermore, a national online survey indicated that the public were willing to pay the tax for reducing the potential environmental risk of pharmaceuticals; the average willingness to pay per household was CHF 100 (EUR 92) per year, generating a total annual economic value of CHF 155 m (EUR 142.6 million) per year.
>
> Source: Summary of case study provided by Florian Thevenon, WaterLex International Secretariat, Switzerland. See OECD (2019) for more information.

For this project, the OECD has extrapolated the costs of the Swiss approach to 28 EU member states, using data on urban population in agglomerations above 80,000 p.e. Depending on the pace of investment, the aggregate level of additional expenditure to implement the Swiss approach to mitigating CECs at EU level (28 member states at the time of drafting) is projected to be between EUR 129 and 206 billion over 2020-2040.

This projection does not ambition to provide an indication of the future costs of treating CECs across the European Union: in practice, while Switzerland was the first country to embark on a national strategy to address CECs in freshwater, other counties are likely to explore a combination of options, potentially reducing investment costs. Technology costs are likely to fall over time as well. Therefore, the projection below can be considered as a theoretical exercise, and provides an upper estimate of the costs of addressing pharmaceutical residues in freshwater in Europe. To minimise these costs, countries may wish to consider a combination of options, as sketched by the OECD below.

Figure 1.4. Total investment needs by 2040 for CEC treatment – extrapolation of the Swiss approach

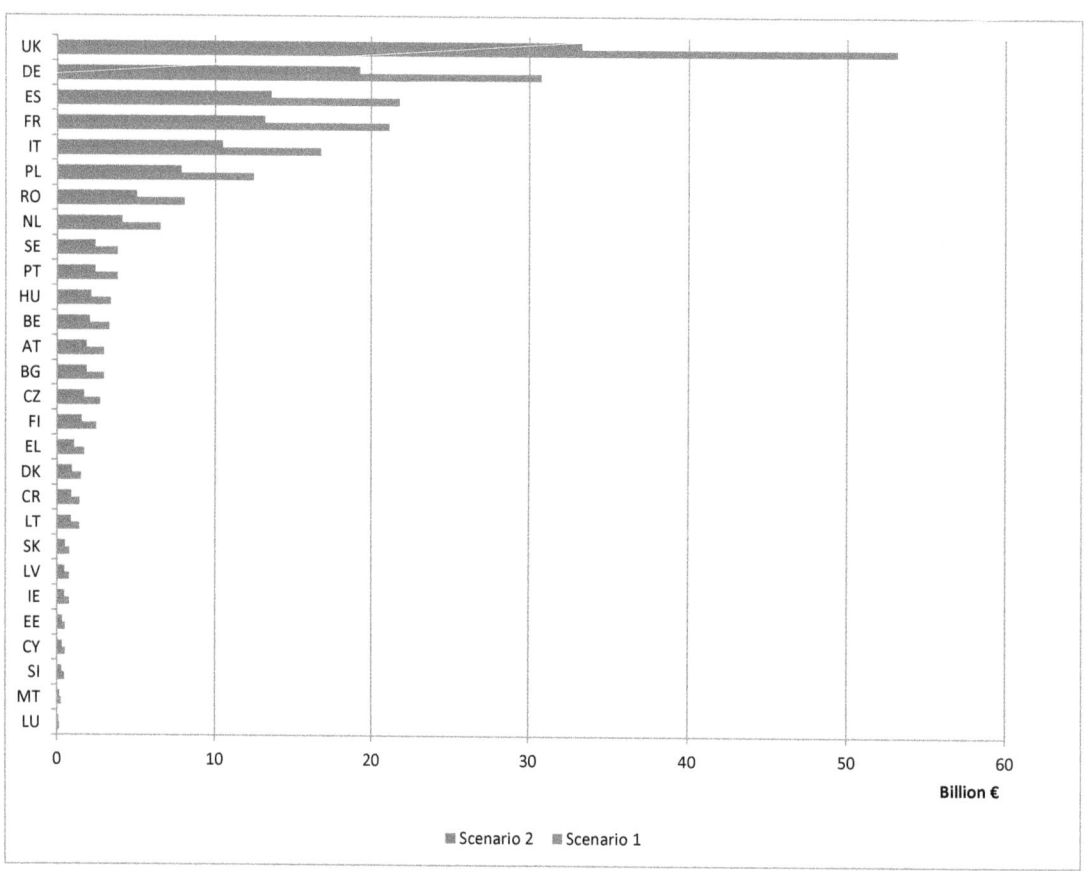

Note: Both scenarios assume that 50% of the population in large agglomerations is connected by 2030 and another 50% by 2040. Scenario 1 assumes a linear investment: each year 0.04% of the population will be connected from 2019 to 2030 and 0.05% between 2030 and 2040. Scenario 2 assumes that 50% of the population is connected in one go in 2030 and another 50% is connected in 2040.
Source: OECD calculation.

From monitoring to taking action: OECD policy recommendations

The policy recommendations were developed independently of the Communication from the European Commission on the same topic: Pharmaceuticals in the environment (EC, 2019b). Both policy guidance documents share a life cycle approach to pharmaceuticals impacts on the environment. The European Commission indicates 6 action areas: actions to raise awareness and promote prudent use, improve training and risk assessment, gather monitoring data, incentivise "green design", reduce emissions from manufacturing, reduce waste and improve wastewater treatment.

Whilst acknowledging that pharmaceuticals will continue to play a necessary and critical role in human and animal health, the OECD identifies five strategies based on proactive policies that can cost-effectively manage pharmaceuticals for the protection of water quality and freshwater ecosystems:

- Improve knowledge, understanding and reporting on the occurrence, fate, toxicity, and human health and ecological risks of pharmaceutical residues in water bodies in order to lay the ground for future pollution reduction measures
- Consider inclusion of environmental risks in the risk-benefit analysis of authorisation of new pharmaceuticals; risk mitigation approaches may be considered for high-risk pharmaceuticals
- Adopt continued environmental monitoring of high-risk pharmaceuticals post-authorisation (including of those already approved on the market)
- Proactively manage pharmaceuticals through drinking water safety plans (see WHO 2011 Guidelines for Drinking Water Quality), monitoring programmes, and incidence reporting to identify and prevent contamination and adapt policy to new science
- Take advantage of alternative innovative monitoring technologies and water quality modelling, to minimise costs. Cost-effective approaches are likely to prioritise substances and water bodies of highest concern, and target areas of high pollution, all of which require improved knowledge.

Source-directed approaches to impose, incentivise or encourage measures in order to prevent their release into water bodies. For example, options may include: i) water quality or technology standards for discharges from pharmaceutical manufacturing plants as part of Good Manufacturing Process audits, product certification and green procurement standards, ii) taxes levied on hazardous substances to incentivise changes in production processes or substitution of substances with less hazardous alternatives (but equally beneficial to human or veterinary health), and iii) subsidies from government to promote and support research and development for green and sustainable pharmacy.

Use-orientated approaches to impose, incentivise or encourage reductions in the inappropriate and excessive consumption of pharmaceuticals. Options may include: i) restrictions on over-the-counter sales of, and self-treatment with, priority pharmaceuticals ii) bans or restrictions on the excessive or unnecessary use of pharmaceuticals that have known harmful environmental impacts (i.e. antibiotics as livestock growth promoters, and diclofenac as pain and inflammation relief), and iii) public health campaigns and training of physicians, pharmacists and veterinarians to promote the rational use (right patient, right drug, right dose, right time) of priority pharmaceuticals.

End-of-pipe measures – as a complimentary measure to the above strategies – that impose, incentivise or encourage improved waste and wastewater treatment to remove pharmaceutical residues after their use or release to the environment. Options may include: i) increased wastewater tariffs or government subsidies to incentivise advanced wastewater treatment plants, ii) best available techniques for improved wastewater treatment at hospitals (if not collected in municipal wastewater treatment system), and iii) extended producer responsibility legislation with regulatory requirements and targets for correct waste disposal of unused pharmaceuticals.

Collaboration and a life-cycle approach. Ultimately, a life-cycle approach combining a policy mix of the above four strategies and involving several policy sectors is required to effectively deal with pharmaceuticals across their life-cycle, including pharmaceutical design, authorisation, manufacturing, prescription, over-the-counter purchases, consumer use (patients and farmers), collection and disposal, and wastewater treatment. Given the risks identified in this report, action should be taken to reduce impacts to the maximum feasible extent throughout the pharmaceutical chain.

All stakeholders along the pharmaceutical chain have a critical role to play in the transition to more effective management of the risks to water quality, ecosystems and human health from pharmaceutical pollution. Voluntary participation alone will not deliver; economic and regulatory drivers from central government are needed.

Policymakers will need to factor in financing measures for the upgrade, operating and maintenance costs of wastewater treatment plants, as well as policy transaction costs to facilitate the transition from reactive to proactive control of pharmaceutical residues in water bodies.

The case of Extended Producers' Responsibility

It is estimated that 10-50% of prescription medications are not taken as per the doctors' orders and unused or expired medicinal waste may be disposed of via the toilet - therefore offering zero therapeutic benefit and resulting in water pollution. Although the contribution of improper disposal of pharmaceuticals to the overall environmental burden is generally believed to be minor (Daughton and Ruhoy, 2009), pharmaceutical collection schemes are still considered to be important (OECD, 2019).

Various systems have been developed around the world to recover and manage waste pharmaceuticals from households. Drug take-back programmes provide the public with a convenient way to safely dispose of leftover medications. In Europe, collection schemes of unused/expired medication are an obligatory post-pharmacy stewardship approach that reduces the discharge of pharmaceuticals into environmental waters (via WWTPs) and minimises the amounts of pharmaceuticals entering landfill sites (OECD, 2019).

Public collection schemes of unused pharmaceuticals are established in several OECD countries either as voluntary schemes or mandated by legislation (Table 1). Collection programmes are funded either by the government (e.g. Sweden, Australia) or by Extended Producer Responsibility (EPR) in line with the Polluters-Pays principle (e.g. in Canada, Belgium, Spain and France) (Barnett-Itzhaki et al., 2016). Through EPR legislation, pharmaceutical companies are required to collect and dispose of the unused pharmaceuticals their companies put on the market. The advantage of EPR systems is that it takes the burden off the government and requires industry to finance and manage the collection and safe disposal (usually through incineration) of unused drugs. Companies can internalise these costs in the price of pharmaceuticals and can, in theory, provide services more cost-efficiently and be incentivised to manufacture more targeted, personalised medicines to avoid wastage. For more on EPR as a policy approach, see (OECD, 2016).

Table 1.3. Household pharmaceutical collection and disposal programmes, select OECD countries

Country	Programme coverage	Method	Funding
United States	28 local EPR laws in the US; 5 at state-level, 23 at local government level	Either voluntary programs by firms or governments, or mandatory programs through EPR	Governmental, Industry
Canada	Several regional programs across the country. Four EPR programs regulated under different jurisdictions	Retail pharmacies commonly act as collection sites	Brand-owners and contributions are based on market share.
Australia	National programme	Mandatory, Retail pharmacies commonly act as collection sites	Federal government
France	National programme	Mandatory EPR-scheme, Retail pharmacies commonly act as collection sites	Industry
Sweden	National programme	Mandatory EPR-scheme, Retail pharmacies commonly act as collection sites	Pharmacies

Source: OECD (2019), Pharmaceutical Residues in Freshwater: Hazards and Policy Responses, OECD Studies on Water, OECD Publishing, Paris, https://dx.doi.org/10.1787/c936f42d-en.

EPR schemes may also be a potential policy option to assist with financing the upgrade of wastewater treatment plants to remove emerging pollutants. In Germany, the central government initiated a national dialogue on emerging pollutants (including pharmaceuticals) in water in 2017. In the face of increasing use of pharmaceuticals causing a rise in pharmaceutical residues in waters, the introduction of advanced (fourth stage) wastewater treatment for agglomerations above 5000 people is being discussed, amongst of other possible policy solutions. As part of the dialogue, the German Association of Energy and Water Industries (BDEW) has proposed an EPR scheme as one way to fund the upgrade of wastewater treatment plants to remove pharmaceuticals (see Box below). The scheme has been specifically discussed in the context of the highly polluted Niers River Basin, which hosts several pharmaceutical manufacturing plants.

Box 1.8. A proposal for an EPR Scheme to recover costs of advanced wastewater treatment plant upgrades, Germany

The cost of upgrading wastewater treatment plants serving a population of >5000 persons in Germany with an advanced (fourth level) of treatment has been estimated to cost €1.2 billion/year or €15.20/person/year. This would result in a wastewater service tariff increase of, on average, 14-17%, and come at a total cost of €36 billion over 30 years.

One financing option proposed is an EPR scheme. Under the proposed EPR scheme, pharmaceutical manufacturing companies operating in a river basin would be obliged to contribute to the cost of wastewater treatment according to their share of pollution (in accordance with the polluter pays principle under the WFD). The EPR scheme is proposed to operate as follows:

- Establishment of a national water fund and coordination unit to manage the scheme
- Wastewater utilities install advanced (fourth treatment) stage at wastewater treatment plants if the following two conditions are realised: i) environmental quality standards (EQS) are exceeded for one or more substances in a water body receiving wastewater discharge (the list of substances with an EQS is expanding, as is water quality monitoring); ii) the polluting companies responsible for the pollution can be identified
- The total costs (capital and O&M costs) of a wastewater treatment plant upgrade are reported to the national water fund coordination unit.
- Each polluting company is obliged to pay for their share of the cost of the wastewater treatment plant upgrades in accordance with the units of pollution emitted each year (determined by a pollution coefficient (indicator of the environmental harm of the polluting substance) and the volume of pollution emitted each year).
- Funds received from polluting companies in the EPR scheme will be distributed to wastewater utilities to refund the cost of advanced treatment.
- The EPR financing option has the following advantages:
- It prioritises wastewater treatment plants for upgrades, based on environmental impacts of harmful polluting substances
- It transfers to the costs of treatment to the polluters, and is therefore in alignment with the polluter pays principle and the WFD
- It provides a financial incentive for polluters to invest in less polluting production processes or more sustainable substances/products (i.e. green pharmacy)
- It is less difficult and has a lower administrative cost than financing by way of a levy (tax) on pharmaceutical products.

However, the proposed EPR scheme would require a legally binding obligation from government for polluting companies to pay.

Source: Civity (2018); personal communication (2019).

Lessons can be learned from on-going discussions in Germany to inform an EU wide reflection on the relevance and feasibility of EPR schemes to address pharmaceutical residues in freshwater systems. A range of options may be considered, including ones based on a simplified approach, inspired by the EPR schemes developed for solid waste management.

Whilst an EPR scheme to finance upgrades in water treatment plants may be more cost-efficient and effective than a simple tax or levy on pharmaceutical products (as outlined in Box 1), it remains that the most long-term, large-scale and cost-effective solutions to reducing pharmaceuticals in the environment is through preventative source-directed and use-orientated policy measures, early in the pharmaceutical life cycle. Such policy measures may include incentives for the design of green pharmaceuticals or personalised medicines, sustainable public procurement with environmental criteria to limit pollution, and improved diagnostics and restrictions on the inappropriate or excessive consumption of pharmaceuticals with high environmental risk (OECD, 2019).

Box 1.9. Policy responses to CECs: A state of flux

To better integrate current and future pollutants emissions, their fate and potential adverse mixture effects, the EU is currently developing a chemical strategy for sustainability in the context of the Green Deal. Recognising that CECs may not be great candidates for classic regulation, the Ministry of Ecological and Solidarity Transition in **France** created a five-year programme with financial incentives (EUR 10 million) aimed at stimulating new innovative projects to manage CECs and empowering local stakeholders. The selected projects targeted domestic, industrial, diffuse and multiple sources of pollution and include solutions for better diagnostics, cost-efficient reduction of CECs and changes in practices of various types of stakeholders.

In **the Netherlands**, a 2015 incident of pyrazole in the River Meuse (an important drinking water source) triggered the development of a water quality standard (WQS) for pyrazole. The incident also led to the creation of a step-by-step action guide for stakeholders to safeguard public health and drinking water production from future CECs pollution events. In addition, the issuance of industrial permits was revised, mandating the inclusion of the potential effects of CECs on drinking water production. They all contribute to the distinctive Chain Approach in the Netherlands.

Most country responses to date have focussed on upgrading wastewater treatment plants. For example, an extensive study by the **Swedish EPA** (2017) of over 450 wastewater treatment plants has confirmed that advanced treatment of pharmaceutical residues in wastewater is necessary given the potential long-term effects to the aquatic environment, anticipation of future regulations, a responsibility to consider the Precautionary Principle, and benefits of being a front runner.

Such a policy comes at a cost. The removal of CECs such as pharmaceuticals and fire retardants in wastewater treatment plants (e.g. by ozonization, active carbon filtration) in **Finland** has been estimated as requiring investments of €700–1400m. This would increase energy use by 30–80% and increase wastewater charges by 9–41%.

In the **United Kingdom**, The UK Chemicals Investigation Programme estimates that the cost of implementing wastewater treatment upgrades to remove pharmaceuticals is GBP 27-31 billion (approximately EUR 32-36 billion) over 20 years.

Source: OECD (2019), Pharmaceutical Residues in Freshwater: Hazards and Policy Responses, OECD Publishing, Paris.

References

Barnett-Itzhaki, Z. et al. (2016), Household medical waste disposal policy in Israel, *Israel Journal of Health Policy Research*, Vol. 5/1, p. 48, http://dx.doi.org/10.1186/s13584-016-0108-1

Brown S., Nicholls R.J., Vafeidis A., Hinkel J., and Watkiss P. (2011), The Impacts and Economic Costs of Sea-Level Rise in Europe and the Costs and Benefits of Adaptation. Summary of Results from the EC RTD ClimateCost Project, in Watkiss P. (Editor), *The Climate Cost Project. Final Report. Volume 1: Europe*, Stockholm Environment Institute, Sweden, ISBN 978-91-86125-35-6.

Cambridge Econometrics (2017), *Bridging the water investment gap*, a report to the European Commission DG Environment

Civity (2018), *Costs of a fourth treatment stage in wastewater treatment plants and financing based on the polluter pays principle* (in German), Civity Management Consultants, Berlin, https://www.bdew.de/media/documents/PI_20181022_Kosten-verursachungsgerechte-Finanzierung-4-Reinigungsstufe-_Klaeranlagen.pdf .

Daughton, C., I. Ruhoy (2009), Environmental footprint of pharmaceuticals: the significance of factors beyond direct excretion to sewers, *Environmental Toxicology and Chemistry*, Vol. 28/12, p. 2495, http://dx.doi.org/10.1897/08-382.1.

Dolowitz D.P., Bell S., Keeley M. (2017), Retrofitting urban drainage infrastructure: green or grey?, *Urban Water Journal* 15(1):1-9, DOI: 10.1080/1573062X.2017.1396352

ECONADAPT (2015), *The Costs and Benefits of Adaptation: Results from the ECONADAPT Project.* Watkiss, P. (ed.), ECONADAPT consortium

EurEau (2017), *Europe's water in figures. An overview of the European drinking water and waste water sectors*, Brussels

EurEau (2016), *Overflows from Collecting Systems*, Brussels

European Commission (2019a), *Fitness Check of the Water Framework Directive and Floods Directive – SWD(2019) 439*, Brussels

European Commission (2019b), *European Union Strategic Approach to Pharmaceuticals in the Environment*, COM(2019) 128 final

European Commission (2017), *Study supporting the revision of the EU Drinking Water Directive*, Luxemburg, doi:10.2779/650074

European Environment Agency (2017), *Climate change, impacts and vulnerability in Europe 2016. An indicator-based report*, Luxemburg, doi:10.2800/534806

European Court of Auditors (2017), *Special report no 12/2017: Implementing the Drinking Water Directive*, Luxemburg

European Investment Bank (2016), *Restoring European Competitiveness*, Luxemburg

Hinkel J., Nicholls R.J., Vafeidis A.T., Tol R.S.J., Avagianou T. (2010), *Assessing risk of and adaptation to sea-level rise in the European Union: An application of DIVA*. Mitig Adapt Strateg Glob Change 15:03-719

IWA (2016), *Performance Indicators for Water Supply Services: Third Edition*, IWA Publishing

Jongman B. et al. (2014), Increasing Stress on Disaster-Risk Finance due to Large Floods, *Nature Climate Change*, 4(4), DOI: 10.1038/NCLIMATE2124

Milieu (2016), *Assessment of impact of storm water overflows from combined waste water collecting systems on water bodies (including the marine environment) in the 28 EU Member States*, Brussels

National Infrastructure Commission (2018), *National Infrastructure Assessment*,

OECD (2019), *Pharmaceutical Residues in Freshwater: Hazards and Policy Responses*, OECD Studies on Water, OECD Publishing, Paris, https://dx.doi.org/10.1787/c936f42d-en.

OECD (2018), *Financing water - Investing in sustainable growth: Policy Perspectives*, OECD Publishing, Paris

OECD (2017), *Diffuse Pollution, Degraded Waters: Emerging Policy Solutions*, OECD Studies on Water, OECD Publishing, Paris. http://dx.doi.org/10.1787/9789264269064-en

OECD (2016), *Extended Producer Responsibility: Updated Guidance for Efficient Waste Management*, OECD Publishing, Paris, https://dx.doi.org/10.1787/9789264256385-en.

OECD (2015), *Climate Change Risks and Adaption: Linking Policy and Economics*, OECD publishing, Paris

OECD (2013a), *Water Security for Better Lives*, OECD Studies on Water, OECD.

OECD (2013b), *Water and Climate Change Adaptation. Policies to Navigate Unchartered Waters*, OECD Publishing, Paris

OECD (2006), *Infrastructure to 2030: Telecom, Land Transport, Water and Electricity*, OECD Publishing, Paris, ISBN: 9264023984

Sadoff et al. (2015), *Securing Water, Sustaining Growth*: Report of the GWP/OECD Task Force on Water Security and Sustainable Growth, Oxford University

Spit W., et al. (2018), *The Economic Value of Water - Water as a Key Resource for Economic Growth in the EU*, deliverable to Task A2 of the BLUE2 project "Study on EU integrated policy assessment for the freshwater and marine environment, on the economic benefits of EU water policy and on the costs of its non- implementation", report to Directorate-General for the Environment of the European Commission

Swedish EPA (2017), *Advanced Wastewater Treatment for Separation and Removal of Pharmaceutical Residues and Other Hazardous Substances. Needs, Technologies and Impacts*, Bromma, Sweden

UKWIR (2017), *Achieving Zero Leakage by 2050: Basic Mechanisms of Bursts and Leakage*, London

USEPA (2016), *Combined Sewer Overflow Management Fact Sheet. Sewer Separation*, Washington DC

WAREG (2017), *An Analysis of Water Efficiency KPIs in WAREG Member Countries*, WAREG

Winpenny J. (2015), *Water: fit to finance? Catalyzing national growth through investment in water security*, report of the High-Level Panel on Financing Infrastructure for a Water-Secure World, World Water Council and OECD

Notes

[1] At the time of the publication of this report, the negotiations for adopting a revised Drinking Water Directive are still ongoing.

[2] As reported in the information provided by member states to the European Commission when reporting on the state of implementation of the Directive according to Article 17.

[3] European Commission (2019a), Fitness Check of the Water Framework Directive and Floods Directive – SWD(2019) 439, Brussels

[4] This corresponds to a combination of Representative Concentration Pathways 8,5 and Shared Socio-economic Pathways 2 from the Intergovernmental Panel on Climate Change 5th Assessment report on climate and socio-economic change scenarios. See the methodological note for more information.

[5] This section builds on OECD (2019), Pharmaceutical residues in freshwater: Hazards and policy responses, OECD Publishing, Paris

2 The state of play

This chapter documents current levels of expenditure for water supply and sanitation, in the European Union. It reports on sources of finance in use (the 3Ts) and affordability issues. It provides a robust benchmark of the state of play across EU member states, which is used as a baseline in subsequent chapters of the report.

The chapter presents data on expenditures for flood protection as well. It analyses data gaps.

This part of the report characterises the state of play: compliance with selected Directives; levels of expenditures; and past sources of finance. It covers separately water supply, sanitation and flood protection. It compares the situation of the 28 EU Member States.

2.1. Water supply and sanitation

2.1.1. Compliance and performance

The situation differs between water supply and sanitation. The Drinking water Directive was issued in November 1998. Based on a detailed assessment (EC, 2017a), it is acknowledged that the Directive has led to compliance with the microbiological and chemical parameters set in the regulation; compliance rates are only below requirements for indicator parameters that are of no direct threat to human health, such as taste and odour. This translates into improved safety for human health. Compliance has remained stable over the past years. Challenges remain to maintain the high level of compliance (which requires proper operation, maintenance and renewal of existing assets) and to adjust to the requirements of the revised Directive (still under discussion).

The primary instrument used in the EU to set the performance of wastewater services is the Urban Wastewater Treatment Directive (UWWTD). It mandates that wastewater from urban areas be collected and conveyed to a place of treatment. Treatment should be commensurate with the receiving waters' capacity to cope with the residual pollutants. Treatment needs reflect compliance with other Directives, which contribute to the quality of freshwater in Europe by controlling emissions (Nitrates, Industrial Emissions Directives) or distinctive water bodies (Directives on groundwater or bathing waters).

In its biennial compliance assessment, the European Commission noted that 95% of the EU's urban waste water is collected with 89% receiving secondary and 85% more stringent treatment in line with the UWWTD requirements (EC, 2017b). Trends in compliance have been positive, with major gains made in the EU-13 since 2009; compliance rates have declined since the previous review from the European Commission, reflecting more accurate reporting. Nevertheless,

- Four member states have compliance rates on collection rates below 70% (Bulgaria, Cyprus, Romania, Slovenia). Collection rates decreased from the 8th Report due to the inclusion of Italy, Poland and more accurate data for Romania. Cyprus and Spain have failed to maintain or improve earlier compliance rates.
- Several member states have very low compliance rates as regards secondary treatment (Bulgaria, Malta, Romania, Slovenia). So does Ireland (because of a specific problem in Dublin in 2014).

As mentioned above, for all member states, an additional challenge is to operate, maintain and renew existing assets (European Commission, 2015), to guarantee service quality over time. The rate of asset renewal is not known with accuracy. When it is documented, it is well below the rate of 2%, which is considered by EurEau as appropriate (assuming that assets for water supply and sanitation need to be renewed every 50 years, on average[1]). There is a paucity of data, confirmed in the Figure below. For countries where data is available, rates of renewal can be as low as 0.5%, indicating that infrastructure would only be renewed every 200 years, by far exceeding the life expectancy of the equipment. This confirms the EIB statement that "much of Europe's vital drinking water supply and wastewater management infrastructure is reaching the end of its economic life" (EIB, 2016, p. 34).

| 47

Figure 2.1. Rate of asset renewal for water supply & Rate of asset renewal for sanitation

Rate of asset renewal for water supply

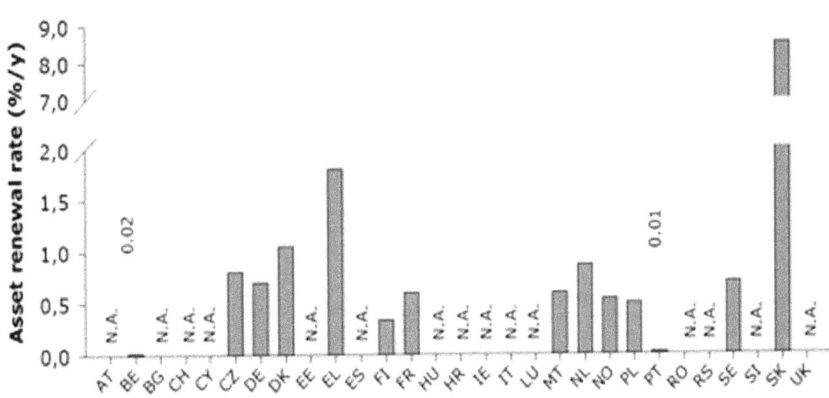

Rate of asset renewal for sanitation

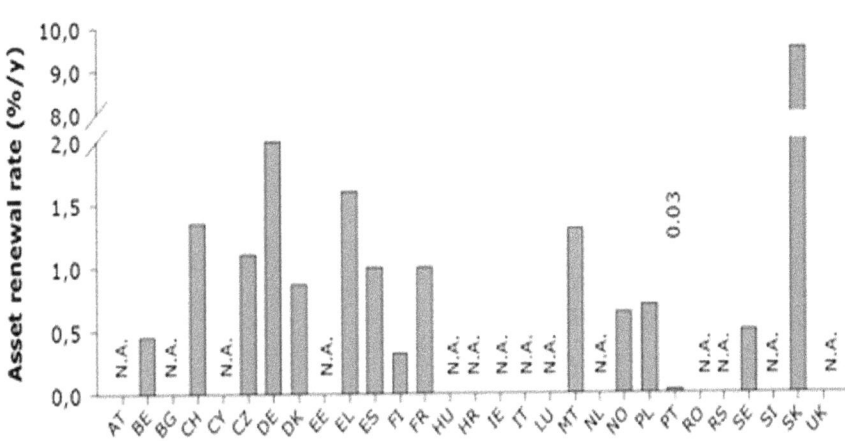

Source: EurEau (2017). Data from 2012-2015, depending on countries.

2.1.2. Expenditure levels

Reference data (annual average for the period 2011-2015) was computed based on a range of Eurostat datasets covering various parts of water-related public and household expenditure (see Annex A for further details). Such data made it possible to establish separate baselines of total expenditures for water supply and sanitation respectively (Figure 2.2). These figures combine both CAPEX and OPEX.

The main limitation of the baseline data used is that it likely over-estimates water supply (and correspondingly underestimates water sanitation) for countries where wastewater-related charges are included in the water bill. Overall, baseline estimates point out to an annual average expenditure of EUR 100 billion across the 28 EU member states, with the lion's share attributable to EU15 (Germany, France, United Kingdom and Italy in particular).

Figure 2.2. Estimated annual expenditures for water supply and sanitation for the EU-28

(million EUR, 2011-15 annual average)

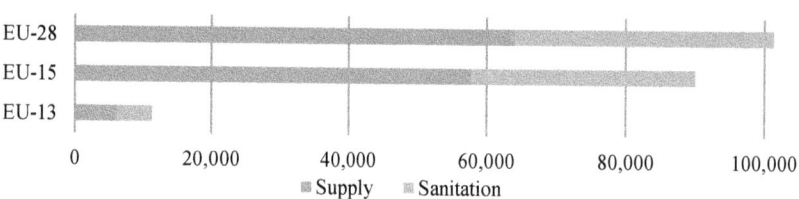

Note: Likely overestimate of supply-related expenditures (and corresponding underestimate of sanitation) in countries where wastewater-related charged are included in the water bill.
Source: EUROSTAT (General government expenditure by function, Final consumption expenditure on environmental protection services by institutional sector, Final consumption expenditure of households by consumption purpose, Mean consumption expenditure by detailed COICOP level).

Figure 2.3. Estimated annual expenditures for water supply and sanitation per member state

(million EUR, 2011-15 annual average)

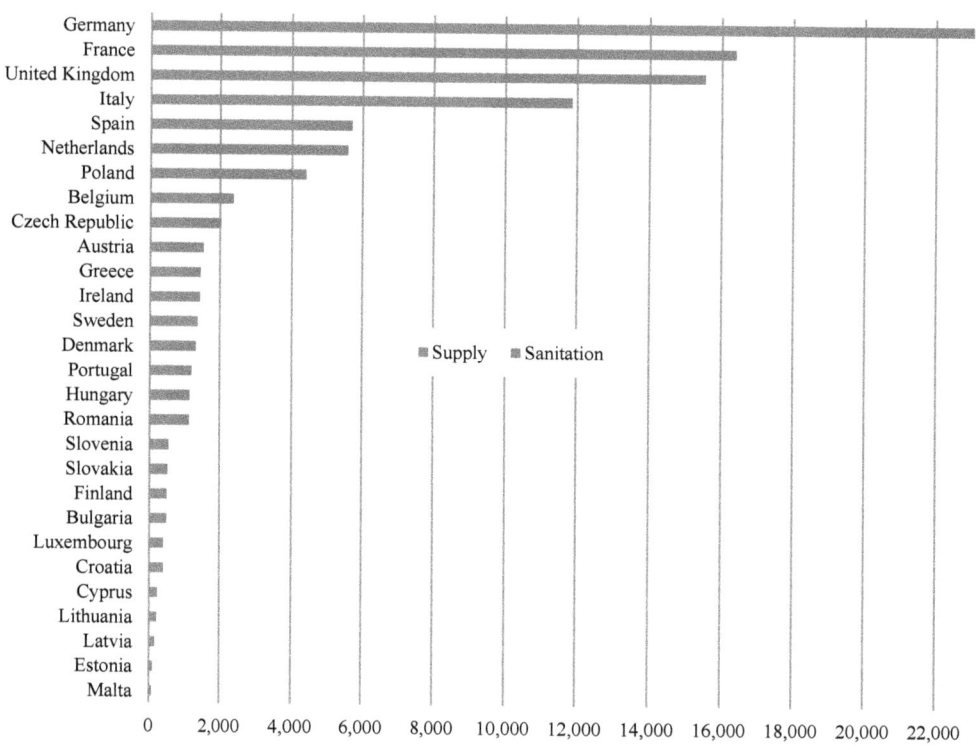

Note: Likely overestimate of supply-related expenditures (and corresponding underestimate of sanitation) in countries where wastewater-related charged are included in the water bill (e.g. Cyprus). Total expenditure for Finland, Croatia and Sweden are known to be underestimated due to data limitations.
Source: EUROSTAT (General government expenditure by function, Final consumption expenditure on environmental protection services by institutional sector, Final consumption expenditure of households by consumption purpose, Mean consumption expenditure by detailed COICOP level).

Figure 2.4. Estimated expenditures per capita for water supply and sanitation in EU-28

(EUR, 2011-15 annual average)

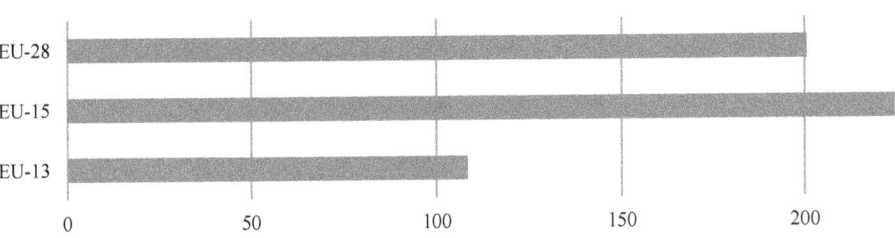

Source: OECD analysis based on EUROSTAT (General government expenditure by function, Final consumption expenditure on environmental protection services by institutional sector, Final consumption expenditure of households by consumption purpose, Mean consumption expenditure by detailed COICOP level).

Figure 2.5. Estimated annual expenditures per capita for water supply and sanitation per member state

(EUR, 2011-15 annual average)

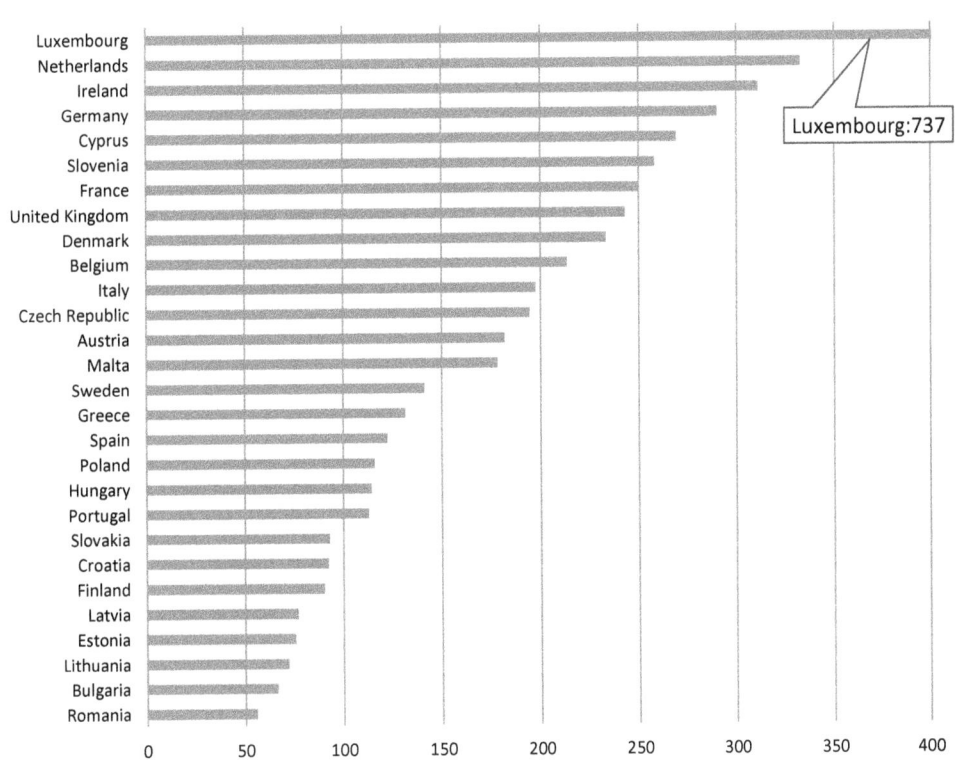

Note: Total expenditure for Finland, Croatia and Sweden are underestimated due to data limitations.
Source: OECD analysis based on EUROSTAT (General government expenditure by function, Final consumption expenditure on environmental protection services by institutional sector, Final consumption expenditure of households by consumption purpose, Mean consumption expenditure by detailed COICOP level).

Figure 2.6. Estimated expenditures per capita and as % of GDP

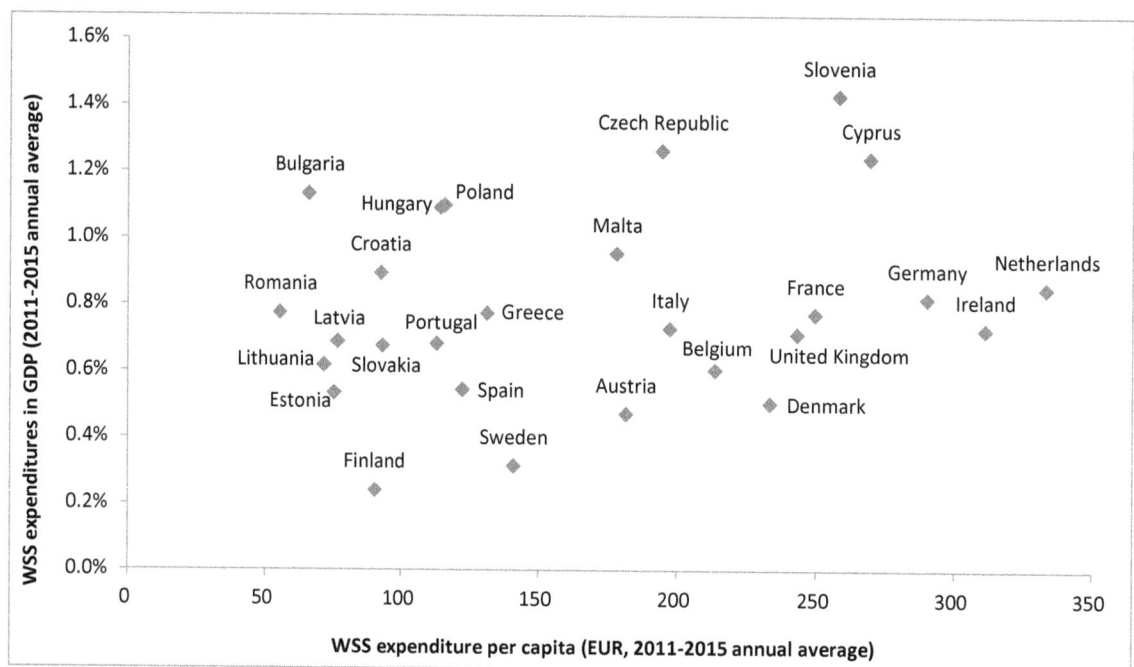

Note: Expenditure for Finland, Croatia and Sweden are underestimated due to data limitations.
Source: OECD analysis based on EUROSTAT (WSS-related public and household expenditures, GDP, population).

For the record, EurEau (2017) reports that, water supply and sanitation utilities invest EUR 45 billion annually in Europe. On average, this represents EUR 93.5 per habitant and per year. This average masks large discrepancies (a factor 10 or more) between member states where utilities invest the least (Czech Republic, Greece, Italy, Malta, Romania, or Slovakia) and countries where they invest the most (Slovenia).

2.1.3. Sources of finance

As aforementioned, estimates of total expenditures in each country were derived from data on current levels of water-related expenditures by the public sector and household respectively. Complementary data sources were then used to estimate the share of total expenditures having relied on EU funding and debt finance. In other words, EU funding and debt are not considered as additional to WSS expenditures but as underlying sources of financing. All data series are further detailed in Annex A.

Figure 2.7 documents the share of the public budgets (consolidated across levels of government) and revenues from water tariffs (i.e. households) in total expenditures. Getting closer to a 100% share of tariffs demonstrates an increasing ability to rely on pricing to finance both capital and operational expenditures. A close to 100% share of public budgets illustrates an absence of pricing. In between the two extremes, a wide range of factors may explain countries' relative positioning on this spectrum.

The data sources cover neither Croatia nor Sweden. An analysis of financial flows in selected countries by EurEau (2018) indicates that in Sweden the costs of service provision are essentially covered by revenues from tariffs. However, taxation is possible according to the legislation; and a few small municipalities subsidise water services through their own local budget. In Croatia, households connected to water supply and sanitation infrastructure typically pay tariffs, in line with the principle of full-cost recovery, although subsidies exist when charges exceed a price cap relative to household income (Danube Water Programme, 2015).

Figure 2.7. Sources of finance for water supply and sanitation services for the EU-28

(2011-15 annual average)

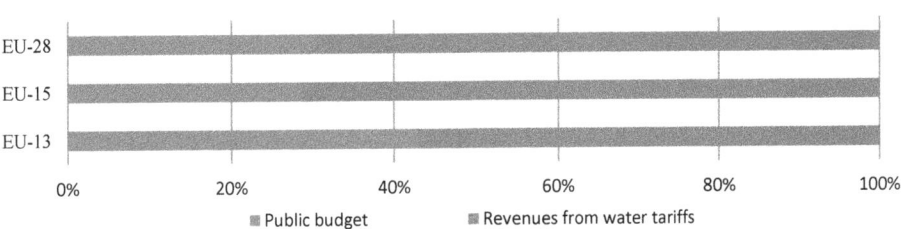

Source: EUROSTAT (General government expenditure by function, Final consumption expenditure on environmental protection services by institutional sector, Final consumption expenditure of households by consumption purpose, Mean consumption expenditure by detailed COICOP level).

Figure 2.8. Sources of finance for water supply and sanitation services per member state

(2011-15 annual average)

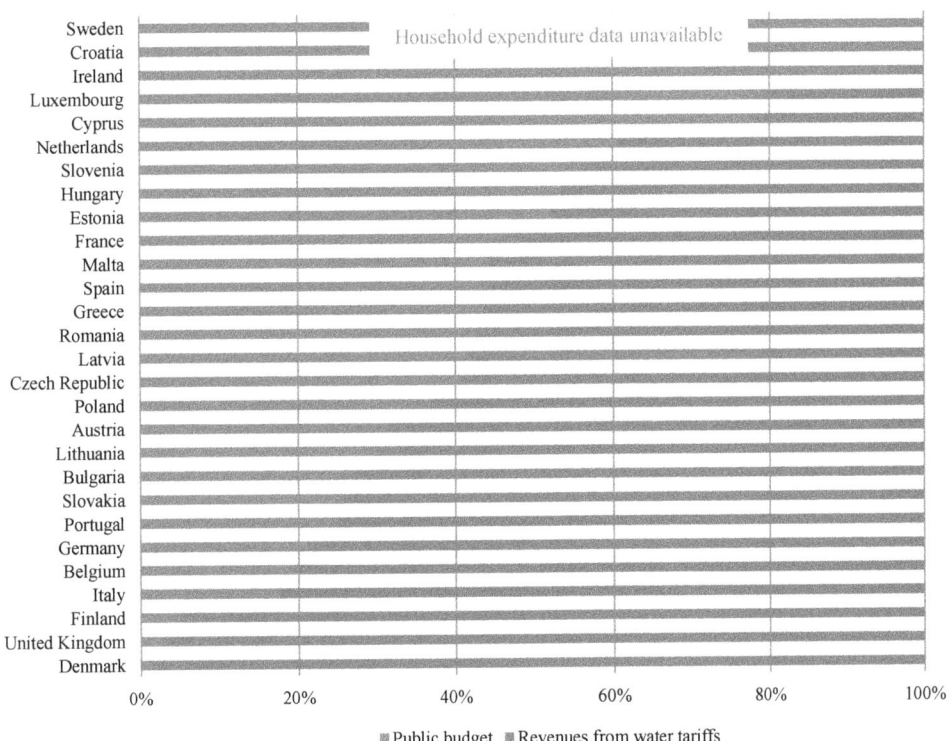

Note: Household expenditures missing for Croatia and Sweden.
Source: EUROSTAT (General government expenditure by function, Final consumption expenditure on environmental protection services by institutional sector, Final consumption expenditure of households by consumption purpose, Mean consumption expenditure by detailed COICOP level).

Figure 2.9 illustrates the extent to which member states' total WSS expenditures (as presented in Figure 2.3 and characterised in Figure 2.8) have been reliant on EU cohesion policy funds (in particular the European Regional Development Fund and the Cohesion Fund) (in most cases provided as grants). The situation varies extensively across countries. In about one third of the Member States, cohesion policy funds have not been extensively allocated to water management over the 2007-2013 and 2014-2020 periods. On the other end of the spectrum, they represent more than 20 % of total expenditure in 8 Member States (and up to 50% in Estonia). National figures may differ from a different study, which covers a longer time period, allowing to factor in expenditures committed 2 year after the end of the financing period (N+2 rule; see COWI, Milieu, 2019).

Figure 2.9. Share of EU funding in estimated total expenditures for water supply and sanitation for the EU-28

(%, 2011-2015 annual average)

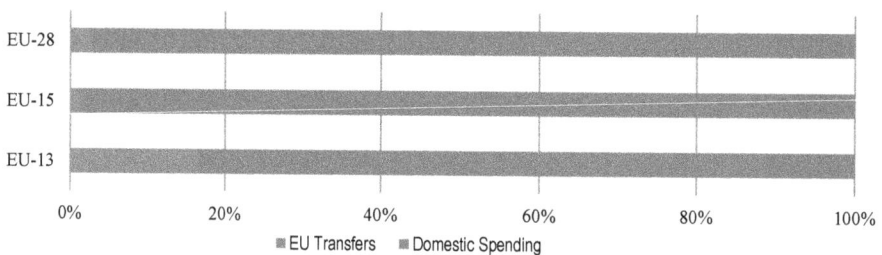

Note: EU cohesion policy funds are channelled through domestic budgets of Member States.
Source: EUROSTAT (for past estimated expenditures), European Commission Directorate-General for Regional and Urban Policy (Open Data Portal for European Structural and Investment Funds).

For a given country (Figure 2.10) a relatively high share of EU funding in total expenditures may result from a number of country characteristics including: the existence of less developed regions, where the bulk of the European Regional Development Fund is invested; access to the Cohesion Fund; an overall low level of expenditures (Figure 2.5 and Figure 2.6), limited domestic financing capacities (see Part IV).

The capacity to apply for funds and spend them effectively matters. In some countries, administrative capacity to design, implement, or operate projects has been the bottleneck: while money was available, projects could not be implemented without delays; or while infrastructures had been built, they remained idle for lack of capacity to operate them.

A complementary analysis of member states' ability to use EU funds effectively was conducted for this project but was inconclusive due to data limitations. Addressing these limitations requires further examination of detailed individual country characteristics, such as institutional arrangements or project pipelines. Of note, in Cyprus, a significant expenditure programme for sanitation was implemented during 2011-2015, funded from the EU cohesion funds; the funding was implemented through the Public Budget.

Figure 2.10. Share of EU funding in estimated total expenditures for water supply and sanitation per member state

(%, 2011-2015 annual average)

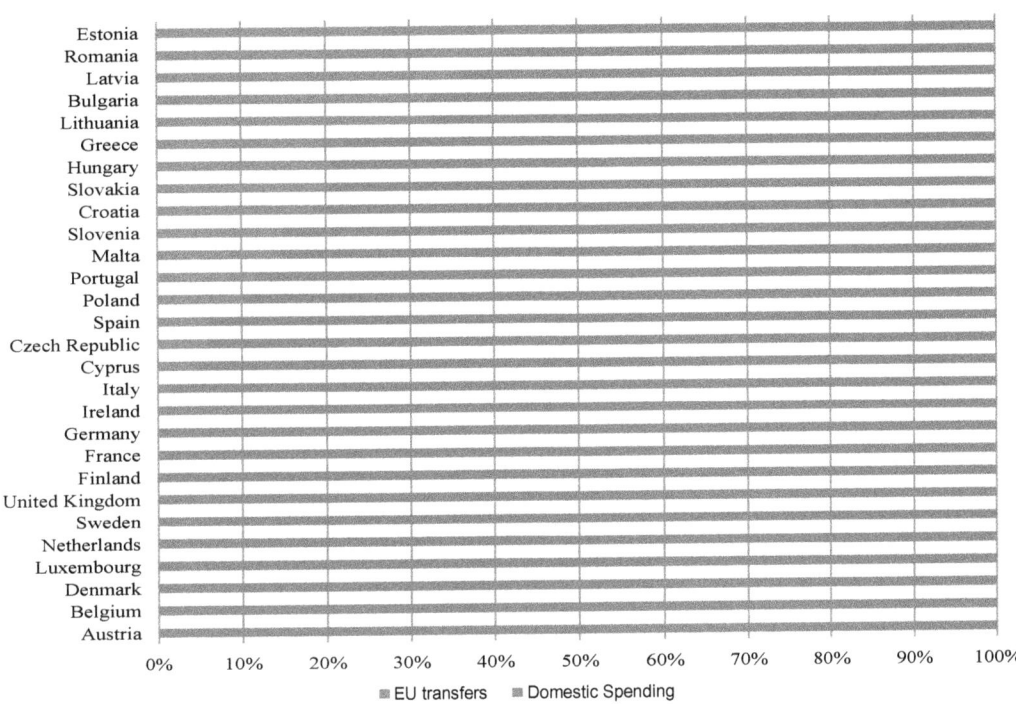

Note: It is assumed that EU funding are always channelled through domestic budgets of each member states and that they are, therefore not additional to government expenditures presented in previous figures.
Source: EUROSTAT (for past estimated expenditures), European Commission Directorate-General for Regional and Urban Policy (Open Data Portal for European Structural and Investment Funds).

Countries may also recourse to debt (reimbursable) finance (Figure 2.11). Doing so typically contributes to financing upfront capital investments where cash flow may be insufficient for on-balance sheet financing or borrowing conditions particularly attractive.

Figure 2.11. Share of debt in estimated total expenditures for water supply and sanitation for the EU-28

(%, 2011-2015 annual average)

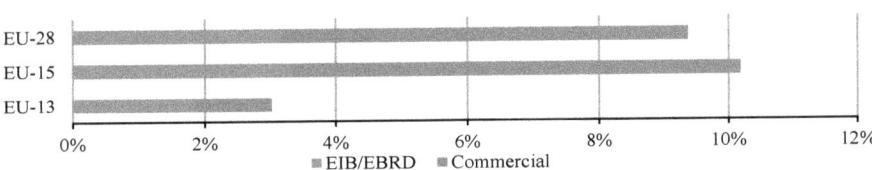

Note: Debt is assumed to be repaid by either (and therefore not additional to) government or household expenditures presented in previous figures
Source: EUROSTAT (for past estimated expenditures), European Investment Bank (loan database), European Bank for Reconstruction and Development (loan database), Commercial databases (IJ Global, Thomson Reuters, Dealogic).

In any given country (Figure 2.12), accessing debt financing will typically be restricted to entities and projects able to demonstrate a reliable ability to pay back. For WSS service providers, such ability is first and foremost dependent on the extent to which costs are recovered through revenues from tariffs or other charges paid by users (see Section 2.1.4 for an analysis of price and affordability; see the Box 2.1 below for a relevant illustration in Sweden). In addition to pricing, the financial health of the entity will greatly influence its ability to access debt finance. For instance, banks are likely to lend to a municipality with low water prices but with a demonstrated ability to raise taxes and featuring a low level of indebtedness (see Part IV for a further analysis of financing capacities).

> **Box 2.1. Financing needs and capacities for WSS in Sweden**
>
> Swedish Investors for Sustainable Development (SISD) is an initiative by SIDA, involving the Swedish Church, AP7, AP3, Skandia and SPP. SISD has focused on investments needed to in Sweden to meet SDG 6. The review indicates that there is more capital than there are investment opportunities within sustainable water and sanitation. The group also came with a series of conclusions, which resonate with the developments in this report:
>
> - The need for investment in water and sanitation infrastructure is large and increasing due to slow renewal rates. Future legal requirements for the disposal of drug residues and micro-plastics can become important factors in investment decisions.
> - New investment tends to override necessary maintenance.
> - Access to funding is available. However, investment is still refrained for several financial reasons.
> - There is an unwillingness to borrow. The municipalities' indebtedness reduces the willingness to invest and, despite the fact that water and sanitation should be kept separate from other municipal activities, these loans are regarded as included in the total debt.
> - The politicians who determine the water tariffs reject raising the tariffs, which also limits the investment rate.
> - Private ownership of water and sanitation infrastructure is not legal in Sweden, which may also restrict private investment.
> - Green bonds are used and will increase in importance. A governmental investigation is due in November.
> - The provision of skills is a major problem. Large demand for consultants, contractors and staff limits the possibilities to reinvest in and maintain existing infrastructure. It is difficult for especially smaller municipalities to manage long-term strategic planning.
>
> Source: AP7 (2017).

Figure 2.12. Share of debt in estimated total expenditures for water supply and sanitation per member state

(%, 2011-2015 annual average)

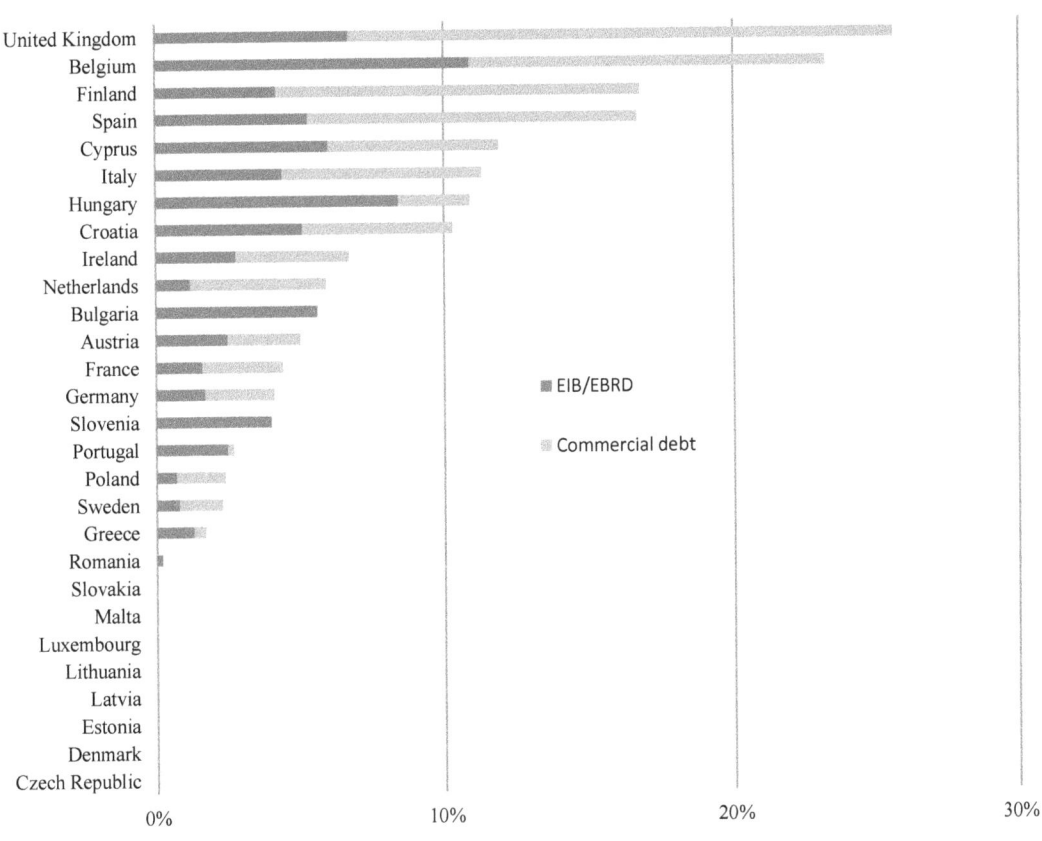

Note: Debt is assumed to be repaid by either (and therefore not additional to) government or household expenditures presented in previous figures.
A study by DANVA (2019) documents debts of water utilities in Denmark. It shows that levels of debt of utilities to KommuneKredit has been rising since 2007 and reached EUR 2 billion at the end of 2017.
Source: EUROSTAT (for past estimated expenditures), European Investment Bank (loan database), European Bank for Reconstruction and Development (loan database), Commercial databases (IJ Global, Thomson Reuters, Dealogic).

2.1.4. Affordability

As mentioned above, pricing is a key element to make WSS services financially sustainable. However, assessment of the second cycle of implementation of river basin management plans indicates that one third of the countries apply cost recovery to a narrow definition of water services, limited to water supply and sanitation services. Compared to the first generation of RBMPs, several countries now apply cost recovery to a wider range of water services, including hydropower generation, navigation, flood protection, or self-abstraction for agriculture and industry. This observation corroborates analyses by WAREG (2017) - the network of water regulators - that full cost recovery through tariffs has not yet been achieved in many EU member states, preventing the sector from being adequately funded.

Affordability concerns, whether perceived or effective, may restrict the ability to use and increase price as a way to recover costs of service provision. In this context, Figure 2.13 plots the share represented by per capita average expenditure for WSS in the lowest 5% and 10% household disposable income[2].

Based on current household expenditure levels, all countries remain below the 3% threshold[3] if considering the lowest quartile and quintile. In a number of countries, shares for the lowest 10% and even more so for the lowest 5% tend to be significantly higher (compared to other EU Member States), which typically reflects a drop in income levels (income inequality).

Figure 2.13. Share of water supply and sanitation expenditures in households' disposable income (2011-2015 average)

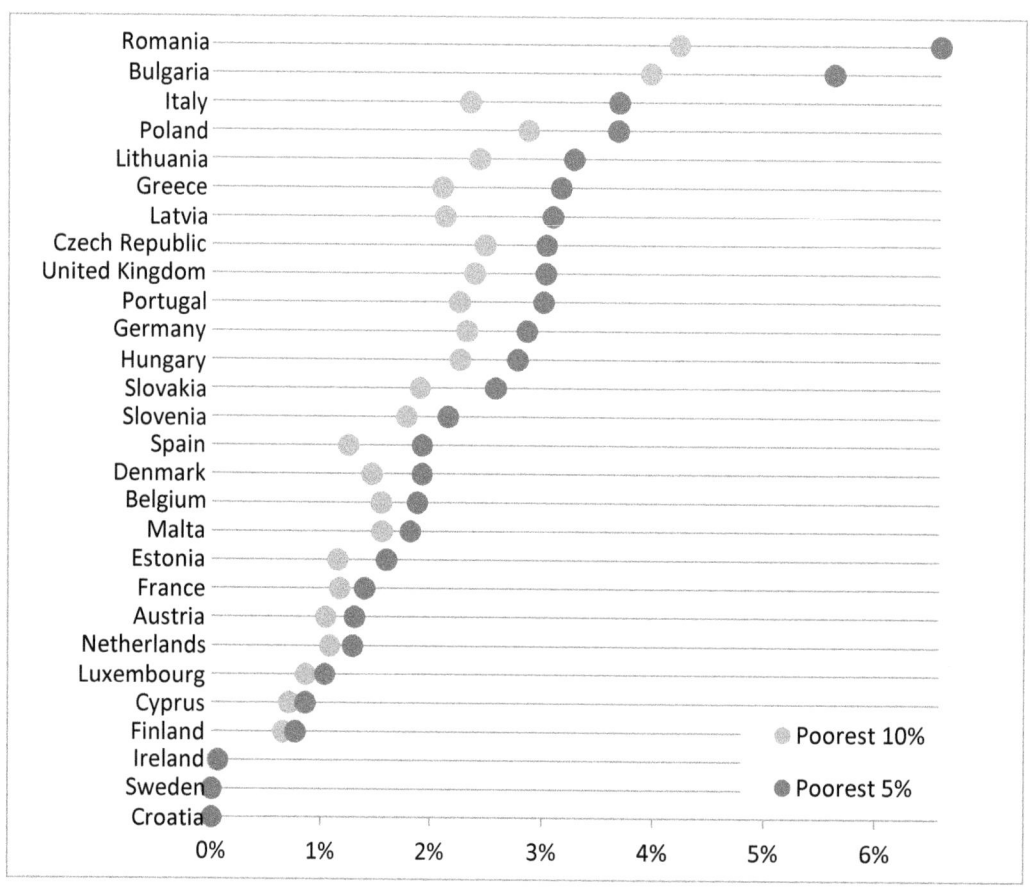

Note: Lack of household expenditure data for Croatia and Sweden.
Source: EUROSTAT (household expenditures and income data).

The data above come with two caveats. First, on-going research (in particular by Guy Hutton or Bob Hope) demonstrates that the statistics above fail to fully capture the complexity of affordability issues. Typically, poorest and most vulnerable households may not pay for public water supply and sanitation, because they are deprived from access to any service. This is typically the case of migrants, homeless, or remote and rural communities.

A second caveat is that estimates presented in Figure 2.13 remain dependent on current level of household expenditures, which in turn very much depends on the extent to which water is actually priced. Hence, affordability issues will be underestimated or may even go unnoticed in countries with a combination of low overall expenditure levels and low to no pricing. On the other hand, countries with reasonably low affordability concerns despite relatively high water prices are in principle in a better position. Section 4.1.1

further develops this discussion by simulating the effect of cover the full cost of WSS through water tariffs (based on current expenditure levels).

2.2. Flood protection

The Floods Directive mandates the development of Flood Risk Management Plans. However, countries vary in their capacity to draft relevant planning documents and their capacity to implement (and finance) them. A recent inventory of first set of Flood Risk Management Plans (FRMPs, 2016) indicates that cost estimates of flood prevention and mitigation measures were reported by about half of the Member States assessed. In many cases, data was only partial (European Commission, 2019a).

In a recent report, the European Court of Auditors signalled that climate change is not properly factored in national plans, as countries rely on past trends to project flood risks. Moreover, implementation of plans can be lax, as plans often lack timelines for implementation and are not accompanied by a financing strategy commensurate with planned expenditures (ECA, 2018).

The Floods Directive does not specify any particular level of security against flood risks. The appropriate level of security remains a political decision, set at country level (OECD, 2013). That decision can be informed by assessments of exposure to risks and of the costs of protection, now and in the future. For instance, in France, the appropriate level of security is set at local level (PAPI) and supported by socio-economic analyses. The point is that no European regulation sets a reference to assess the position of countries as regards performance on flood protection.

Levels of awareness and engagement to mitigate flood risks vary significantly across EU member states. This reflects levels of exposure and experience with flood risks.

2.2.1. Expenditure levels

As further detailed in Annex A, it was not possible to establish a robust baseline of current expenditures, as flood protection does not correspond to a sector or subsector in any international statistical standards/ international classifications. Further, survey data reported by member states are very patchy and unequal (European Commission, 2017c).

Data on projected costs of measures to mitigate flood risks are also incomplete and do not warrant cross-country comparison. The European Commission notes that less than half of member states reported the costs of measures in the first flood risk management plans, and where they exists, reports are incomplete (European Commission, 2019b). Based on that information, the total amount of public finance dedicated to flood protection amounts to EUR 3,070 million per year for the period 2011-15, on average; 3/5 originate in EU 15 countries (Figure 2.14). Note that the order of magnitude matches with the estimate in the Fitness Check: "According to the costs of measures reported in the FRMPs, Member States should invest upwards of €12.5 billion between 2016 and 2021" (European Commission, 2019a, page 57).

Some countries were able to monetise expenditure needs for flood protection in the second generation of river basin management plans. While data in the second generation of plans is a marked improvement over the previous generation, it is not comprehensive enough to support cross-country comparisons: i) not every country has provided such projections; ii) it is not clear whether the countries have used similar definitions and methods.

Figure 2.14. Estimated public budget expenditure for flood protection for the EU-28

(million EUR, 2011-15 annual average)

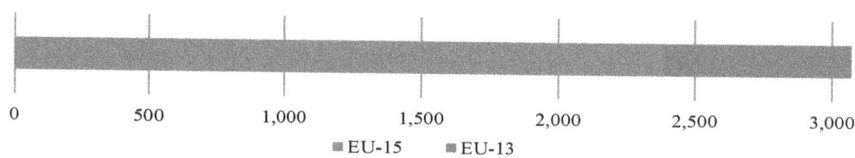

Note: Estimates for most countries are partial to very partial.
Source: European Commission (Flood Risk Management in the EU and the Floods Directive's 1st Cycle of Implementation (2009-15), A questionnaire based report), European Commission Directorate-General for Regional and Urban Policy (Open Data Portal for European Structural and Investment Funds).

Only very few member states reported robust data, which are essentially those most exposed, which invested heavily in flood protection. The only data available relates to public budget expenditures with no records of possible household or other private expenditures. A qualitative overview of potential innovative financing mechanisms and sources can be found in Chapter 5.4.

Reported cost data for the Flood Directive in the Member State compliance assessment reports shows high variability. As an illustration, the Fitness Check calculated average capital costs per inhabitant, and those vary from EUR 0.2 in Estonia, to EUR 261 in Slovenia (European Commission, 2019a, pp.116-7).

Figure 2.15. Estimated public budget expenditure for flood protection per member state

(million EUR, 2011-15 annual average)

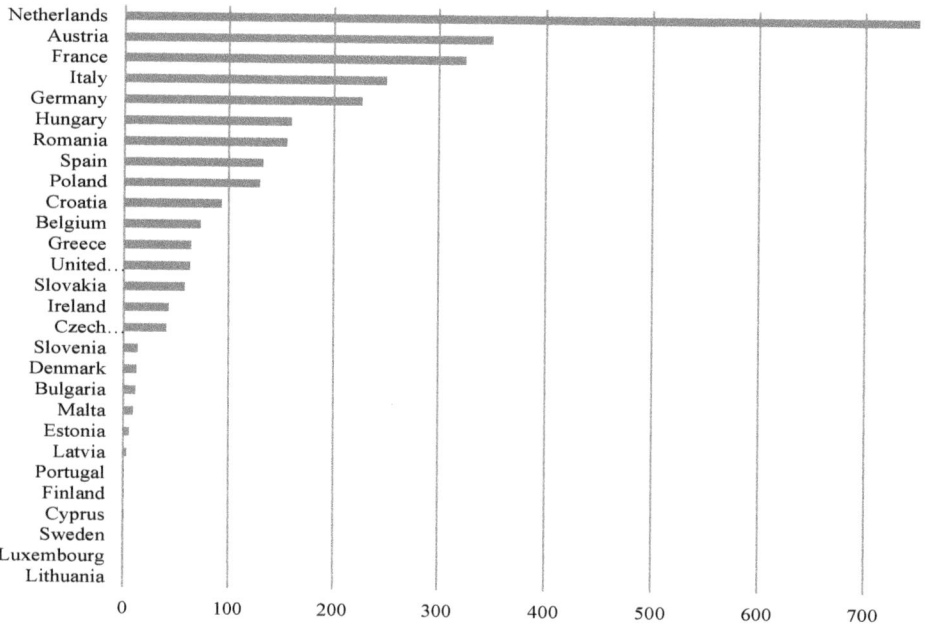

Note: Estimates for most countries are partial to very partial, except for the Netherlands and Austria. Estimates of public expenditures are limited to EU transfer data for the following countries: Bulgaria, Cyprus, Greece, Hungary, Lithuania, Poland, and Slovenia
Source: European Commission (Flood Risk Management in the EU and the Floods Directive's 1st Cycle of Implementation (2009-15), A questionnaire based report), European Commission Directorate-General for Regional and Urban Policy (Open Data Portal for European Structural and Investment Funds).

2.2.2. Sources of financing

Similarly to WSS in the previous section, Figures 2.16 and 2.17 display the share represented by EU funding in the domestic public budget expenditures for flood protection for each member state. In EU-13 countries, that share amounts to 3/5 of the total expenditure for flood protection.

The FRMPs are not a robust source of information to document sources of finance. Funding sources are mentioned, but more often as potential funding mechanisms than as budgetary commitments.

Figure 2.16. Share of EU funding in public budget expenditures for flood protection for the EU-28

(%, 2011-2015 annual average)

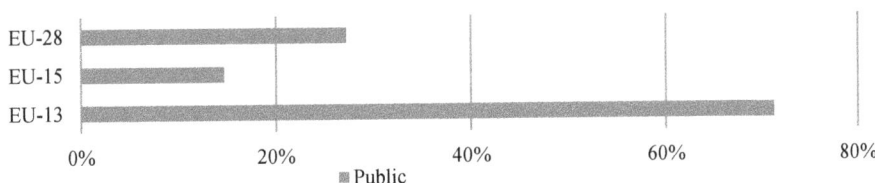

Note: It is assumed that EU funding are always channelled through domestic budgets of each member state and that they are therefore not additional to government expenditures presented in previous figures. 100% reliance on EU funding is in some cases due to the absence of or very limited data on domestic flood expenditures (see previous figure)
Source: EUROSTAT (for past estimated expenditures), European Commission Directorate-General for Regional and Urban Policy (Open Data Portal for European Structural and Investment Funds).

Given the unreliability of expenditure data, these shares should, however, be interpreted with extreme caution. For instance, the fact that one country in three exhibit shares of 100% is, for some, due to the absence of comprehensive domestic public expenditure data, i.e. the only robust data was EU funding.

Figure 2.17. Share of EU funding in public budget expenditures for flood protection per member state

(%, 2011-2015 annual average)

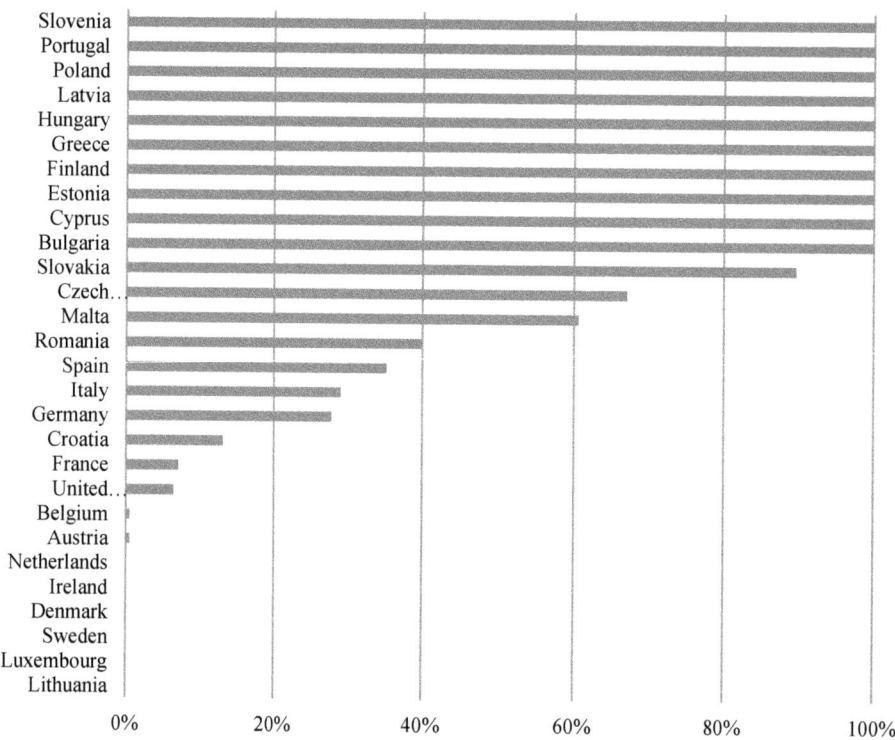

Note: It is assumed that EU funding are always channelled through domestic budgets of each member states and that they are, therefore not additional to government expenditures presented in previous figures. 100% reliance on EU funding is in some cases due to the absence of (for Bulgaria, Cyprus, Greece, Hungary, Lithuania, Poland, and Slovenia) or very limited data on domestic flood expenditures (see previous figure)
Source: EUROSTAT (for past estimated expenditures), European Commission Directorate-General for Regional and Urban Policy (Open Data Portal for European Structural and Investment Funds).

References

AP7 (2017), *Förstudie: Vatten som investeringsobjekt* (with an English Foreword)

COWI, Milieu (2019), Integration of environmental concerns in Cohesion Policy Funds (ERDF, ESF, CF). Results, evolution and trends through three programming periods (2000-2006, 2007-2013, 2014-2020). Final report, European Commission, Brussels

Danube Water Program (2015), Water and Wastewater Services in the Danube Region: A State of the Sector. Regional Report. Danube Water Program. World Bank, Washington, DC.

DANVA (2019), Water in Figures, Danva Statistics & Benchmarking

EurEau (2018), Briefing note. Update on the 3Ts, Brussels

EurEau (2017), Europe's water in figures. An overview of the European drinking water and waste water sectors, Brussels

European Commission (2019a), Fitness Check of the Water Framework Directive and the Floods Directive, Brussels

European Commission (2019b), Implementation of the Water Framework Directive (2000/60/EC) and the Floods Directive (2007/60/EC). Second River Basin Management Plans. First Flood Risk Management Plans, Brussels

European Commission (2017a), Study supporting the revision of the EU Drinking Water Directive, Luxemburg, doi:10.2779/650074

European Commission (2017b), Ninth Report on the implementation status and the programmes for implementation of Council Directive 91/271/EEC concerning urban waste water treatment {SWD(2017) 445 final}

European Commission (2017c), Flood Risk Management in the EU and the Floods Directive's 1st Cycle of Implementation (2009-15). A questionnaire based report.

European Court of Auditors (2018), Special report no 25/2018: Floods Directive, Luxemburg

European Investment Bank (2016), Restoring European Competitiveness, Luxemburg

OECD (2020), Addressing the Social Consequences of Water Tariffs, OECD Working Papers, Paris

OECD (2013), Water Security for Better Lives, OECD Studies on Water, OECD

WAREG (2017), An Analysis of Water Efficiency KPIs in WAREG Member Countries, WAREG

Notes

[1] That figure is indicative only and cannot be taken as a norm or standard. The average masks differences across assets. In principle, most pipes and networks have longer life expectancy than pumps and treatment facilities.

[2] It should be noted that analysis of affordability issues in this report do not rely on tariffs for water supply and sanitation services, but on data on households expenditures. Country-level average prices for water supply and sanitation are not accurate and cannot support robust cross-country comparisons.

[3] It is worth noting that, while 3% is used as a proxy for affordability limit, such a threshold is highly debatable. The level does not build on any robust assessment. The very concept of a threshold is being challenged. See work by Hutton, Wittington. See also OECD Working Paper (forthcoming 2020) *The Social Consequences of Pricing Water*, for a discussion.

3 Projected investment needs across member states

This chapter presents projected financing needs for water supply, sanitation and flood protection, across EU member states. As regards water supply and sanitation, it considers three scenarios: business as usual (where needs are essentially driven by urbanisation), compliance (where countries accelerate efforts to comply with EU Directives, if not already achieved), and efficiency (where countries converge towards arbitrary-set levels of performance for water supply and sanitation services).

Future investments for flood protection are projected but not monetised, as data paucity prevents the construction of a robust baseline.

This part of the report presents projections on financing needs for water supply, sanitation and flood protection for 28 member states by 2050. It ends with a section on issues related to the Water Framework Directive, which could not be quantified, but which affect the volume and nature of investment needs, now and in the future.

Comparisons between different sources for projections are uneasy because of differences in definitions, scope and assumptions. However, aggregate projections on investment needs for water supply and sanitation in the literature are reported, for the record:

- EIB (2016) reports that average annual investment in Europe in 2007-13 in municipal and industrial water and wastewater totalled about EUR 30 billion. The baseline reported in the previous sections amounts to EUR 101 billion; it includes total expenditures including operation and maintenance (not captured by EIB data).
- Projections by GWI indicate a small increase. Average yearly investment could reach EUR 33 billion by 2020. This would not compensate for the current investment backlog.
- EIB (2016) projects that actual investment needs to upgrade and renew Europe's water and wastewater systems are estimated at EUR 75 billion a year for the period 2014-2020. An additional EUR 15 billion would be required to comply with WFD requirements.

3.1. Water supply and sanitation

3.1.1. Business as usual scenario

As noted above, business-as-usual (BAU) projections reflect the additional cost of connecting new city dwellers: they are driven by urban dynamics. The projections do not take account of the rate of use of installed capacity. This may result in projections being quite accurate for Ireland (where installed capacity is fully used in Dublin) but being overestimates of financing needs in Germany or Lithuania (where installed capacity is sufficient to service more city dwellers).

The charts below reflect total additional investment needs for the baseline and business as usual scenario between now and 2030, per country and per capita respectively. It makes no assumption as to how this total amount can be spread annually over the period. Aggregate figure for the 28 member states amount to EUR 1,692 billion.

Figure 3.1. Additional expenditures by 2030 for water supply and sanitation
Baseline + Business as usual scenario

Total cumulative expenditures by 2020-2030: Baseline + BAU (EUR bn)

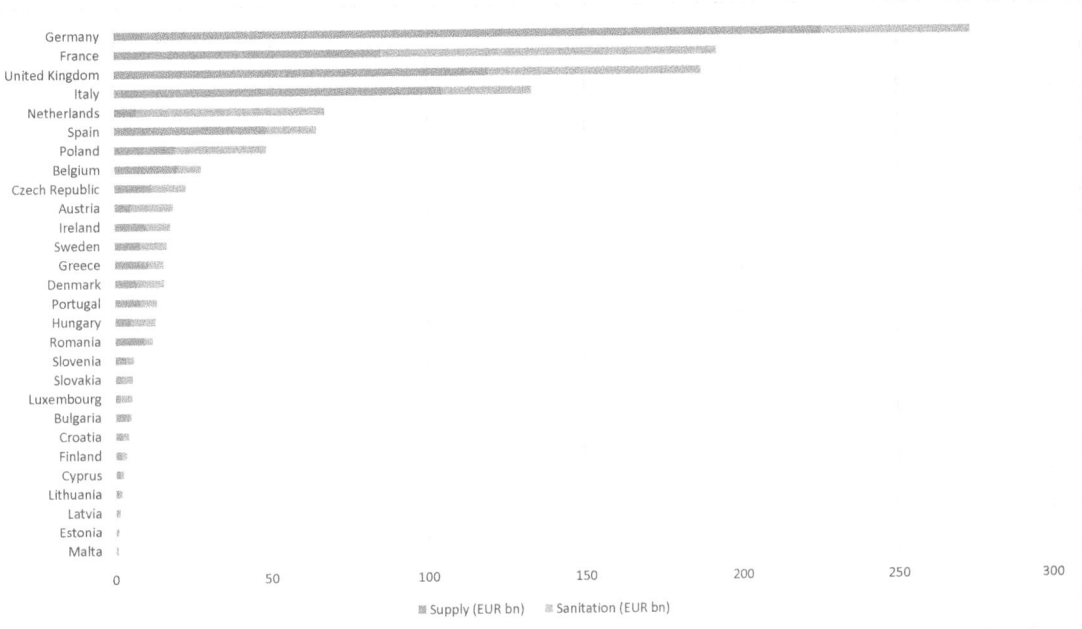

Per capita total cumulative additional expenditures by 2030: BAU (EUR)

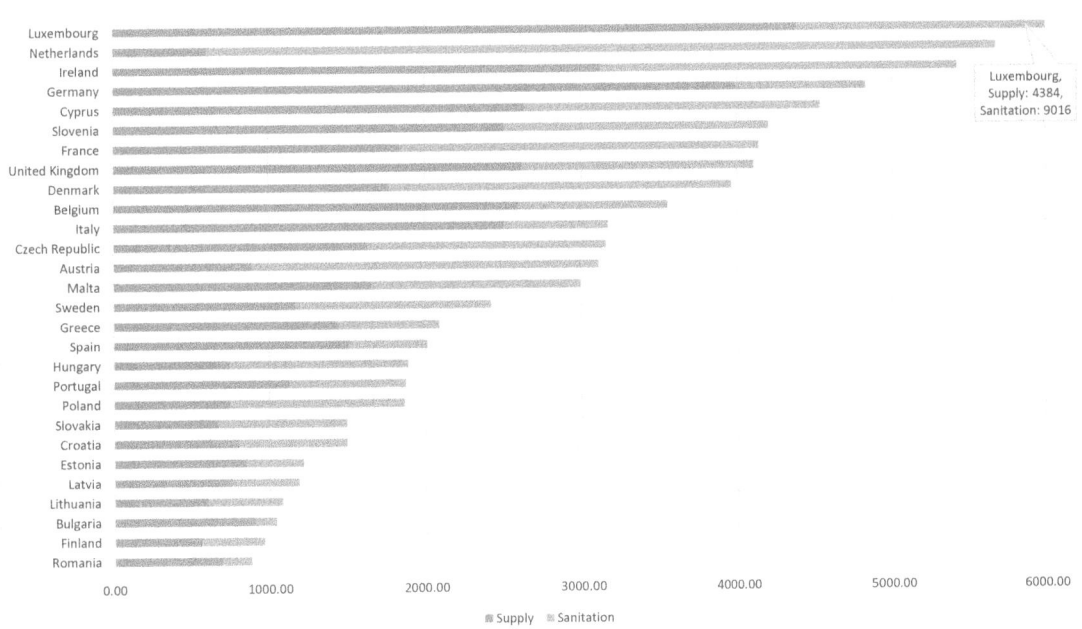

Source: OECD analysis based on EUROSTAT (water-related public and household expenditure data), United Nations and Eurostat (total and urban population statistics and projections).

Nil additional expenditures reflect either no or negative urban population growth. Because fixed costs are an essential part of expenditures for water supply and sanitation, shrinking cities do not enjoy negative additional expenditure over the period, as they still need to operate and maintain existing assets. On the contrary, shrinking urban populations can generate difficulties (and costs) for maintaining WSS assets.

The BAU projections reflect contrasted situations across member states. They do not reflect any potential backlog (or overinvestment) in water supply and sanitation. Over the long term, such a backlog (or overinvestment) translates into the performance of the network. However, it is only partially reflected in leakage rates and non-revenue water. Ideally, projections should be based on a robust knowledge of the state of the infrastructure and history of past investment. Such knowledge however does not exist in most countries; WAREG (2017) notes that poor infrastructure knowledge is a barrier to investment, as it makes it difficult to measure critical issues and to properly plan investments. The Box 3.1 below reports developments that can contribute to more accurate knowledge in the future.

> **Box 3.1. Towards accurate knowledge of water supply and sanitation assets**
>
> France and Portugal have embarked in programmes that can contribute to better knowledge of the state of the assets for water services, thus supporting more accurate planning and decisions for operation, maintenance and renewal.
>
> In France, a regulation issued in 2020 mandates local authorities to inventory public networks for water supply and sanitation. An index was set, to assess compliance with this requirement. When an authority scores below 40 (out of a maximum score of 120), the abstraction charge aide to the Water agency is multiplied by two. There is no such incentive for sanitation. In 2014, 2/3 of water services in France failed to comply with this regulation (figure provided by *Canalisateurs de France*, based on SISPEA data).
>
> In Portugal, ERSAR has developed and is pilot-testing a set of indicators on infrastructure value, infrastructure knowledge and infrastructure management.

3.1.2. Alternative scenario – water supply

An alternative scenario for water supply reflects the cost of compliance with the revised DWD (under discussion at the time of drafting the report) and additional efforts to enhance the efficiency of services. The proxy used for the latter is convergence towards 10% leakage. The Box below discusses this assumption. The scenario adds the cost of connecting vulnerable groups as well. The proposal for the revision of the DWD requires that Member States provide access to water for the vulnerable and marginalised groups.

Urban patterns affect the cost of connecting communities and the efficiency of networks. Sprawl and less dense urbanisation increase the length of networks per user, the risk of leakage and operation and maintenance costs. Additional research could usefully characterise urban patterns for each member state and quantify how they affect future costs and expenditure needs.

Box 3.2. An arbitrary threshold for leakage reduction

The scenario used for projection of investment needs assumes that water utilities in Europe converge towards 10% leakage. This figure is arbitrary. It reflects the best performance of EU countries. It may not be relevant in any context. For instance, leakage is less of an issue where water is abundant and the opportunity cost of using water is low. The appropriate concept is the economic level of leakage, which can only be computed on a case-by-case basis. It is worth noting that leakage wastes more than water: it also wastes energy and other substances used to treat water. A dedicated EU Reference document discusses sustainable levels of leakage (see EC, 2015).

The OECD has tested another threshold for water use efficiency in water utilities in Europe. 20% can be considered a reasonable level of ambition. Several countries are already performing better. These countries would not face additional expenditure to increase efficiency. At aggregate level, relaxing the water efficiency threshold from 10 to 20% would lower investment needs related to water supply under the Compliance and Efficiency scenario by 15%. The Figure below compares the investment needed to achieve both threshold in each member state.

Figure 3.2. Investment needs for water use efficiency per member state

Comparison for 2 levels of efficiency: 10% vs 20%.

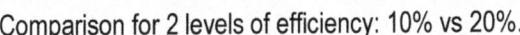

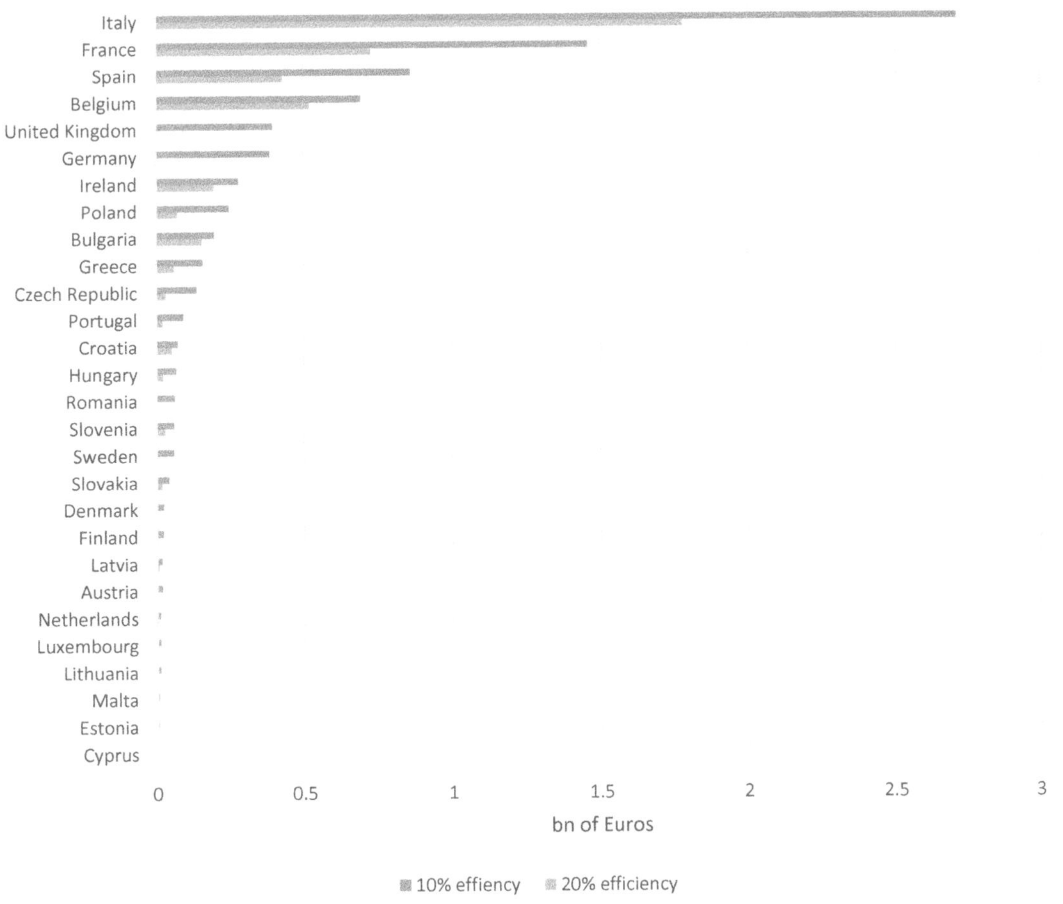

Source: Authors.

Aggregate figures for the 28 member states amount to EUR 35.8 billion. The projections suggest that both the level of additional efforts and the main driver vary across countries. In Romania, the cost of supplying vulnerable groups is disproportionally high. The same situation prevails – to a lesser extent - in Croatia, Poland, Slovakia and the Baltic states. In Italy, efficiency is projected to be a distinctively significant driver for additional expenditures. Belgium, France, Spain and – to a lesser extent – Bulgaria and Ireland face a similar challenge.

Per capita additional levels of effort show a different hierarchy. Romania stands in a distinct category. The atypical ranking of Luxemburg reflects the distinctively high share of non-resident labour, which uses water in Luxemburg during work hours but lives abroad.

Figure 3.3. Additional expenditures by 2030 for water supply - Compliance & efficiency scenario

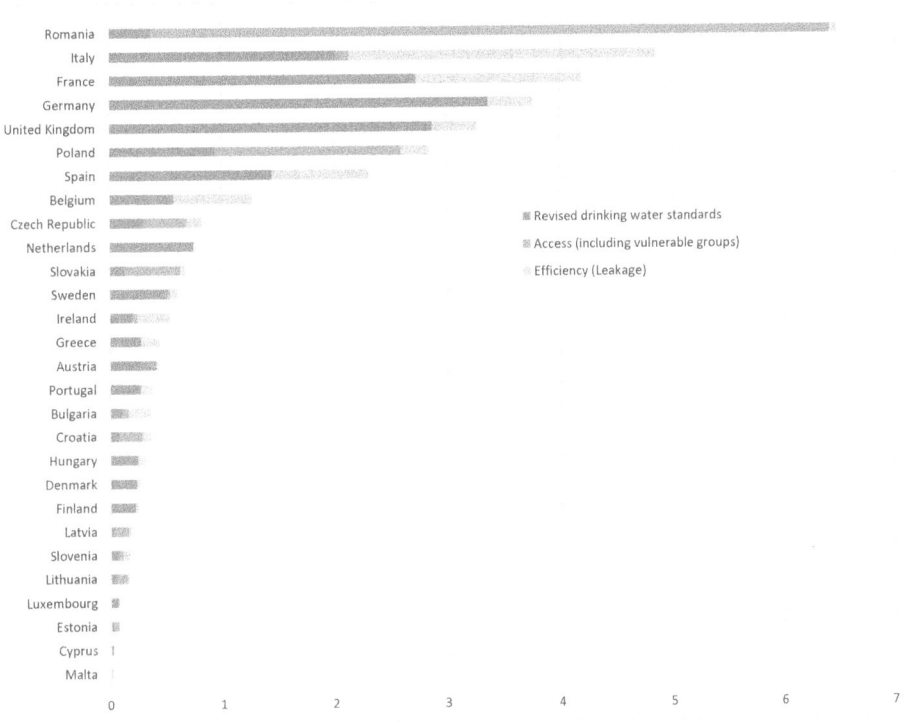

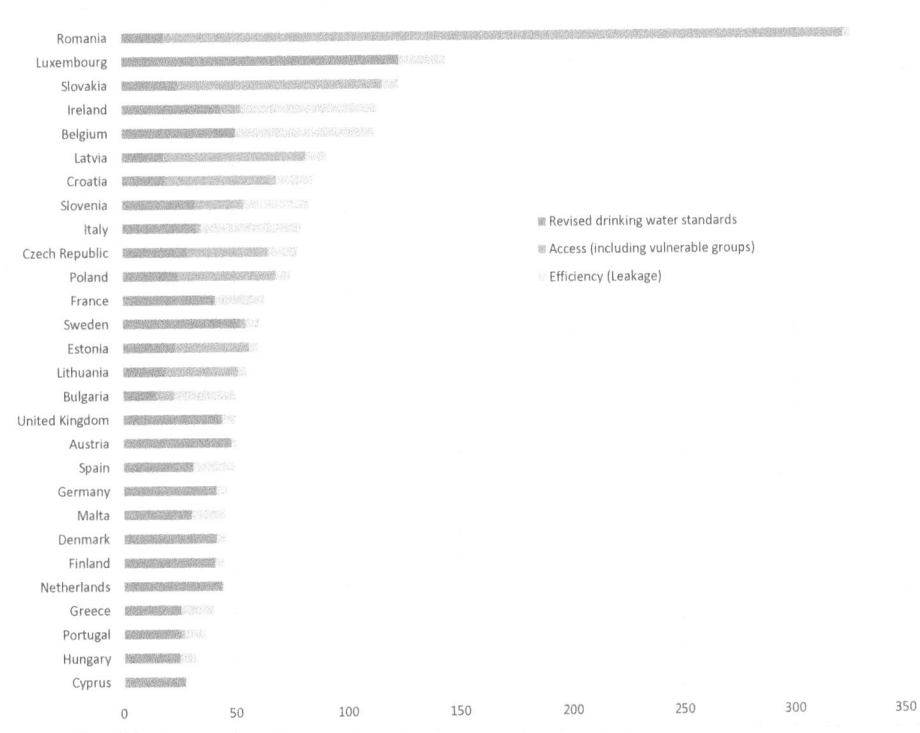

Source: OECD analysis based on European Commission (estimates of costs of compliance with revised DWD, of connecting vulnerable groups, leakage rates) and Eurostat (population) and Eurostat (water-related public and household expenditures).

3.1.3. Alternative scenario – sanitation

An alternative scenario for sanitation captures the additional level of effort required to comply with the UWWTD. The projections are based on the distance to compliance with three key articles of the UWWTD: number of people who need to be connected to a sewer; number of people whose wastewater needs to be treated to a secondary treatment; number of people whose wastewater needs to be treated to more stringent treatment requirements.

Aggregate figure for the 28 member states amount to EUR 253 billion. Ranking of countries according to projected needs does not reflect the EU 15 – 13 categories: Italy, Portugal and Spain still need to invest significantly to comply with the UWWTD. Per capita, Romania and Bulgaria face a distinctively high level of additional expenditures.

The distance to compliance is affected by countries' reliance on individual and other appropriate sanitation systems (IAS; for instance, sceptic tanks). The UWWTD acknowledges that IAS can be appropriate, to avoid unnecessary costs to connect to a centralised collection system. When reporting on distance to compliance, countries assume that IAS comply with UWWTD requirements. This is only the case where IAS are properly designed, their performance is monitored, and compliance is enforced; all conditions which can only be checked on a case-by-case basis.

The European Commission notes that several countries (including Greece, Hungary, Slovakia) report comparatively high levels of reliance on IAS. In selected countries (Czech Republic, Greece, Hungary, Latvia, Poland, Slovakia, Slovenia), IAS collect more than 5% of the total pollution load in agglomerations covered by the UWWTD (2014 data). Greece, Hungary and Latvia feature among the countries which report the smallest distance to compliance, assuming that IAS deliver services in line with UWWTD requirements. The cost of converging towards an arbitrary level of 5% of IAS was computed separately.

Box 3.3. Additional expenditures to converge towards 5% IAS per country

We explore separately integrating IAS levels in relation to distance to compliance. All countries are assumed to have a 5% level of IAS except for the following:

Table 3.1. Converging towards 5% IAS per country

Country	Level of IAS (% of total load)	Additional expenditure to converge towards 5% reliance on IAS (billion EUR; %)	
Slovakia	16.5%	0.598	35%
Hungary	12.7%	1.282	43%
Greece	10.4%	1.086	34%
Poland	8.7%	9.335	71%
Czech Republic	6.8%	0.357	11%
Slovenia	6.2%	.009	1%
Latvia	5.2%	.001	0%
All other countries	5% or below	0	

Note: % is the additional cost compared to the projected cost of Article 3 Compliance only
Source: Level of IAS: European Commission (2014). Additional costs: authors
The analysis of IAS levels examines the additional expenditure required to meet a target IAS of 5%. To achieve this, the estimated expenditures required to connect 95% of the population was to the initial distance to compliance calculation.

Country workshops have signalled situations where central systems are in place, but dwellers are reluctant to connect, because they do not want to pay - or cannot afford - the cost of connection. Lithuania is an example.

Figure 3.4. Additional expenditure by 2030 for sanitation – Compliance scenario

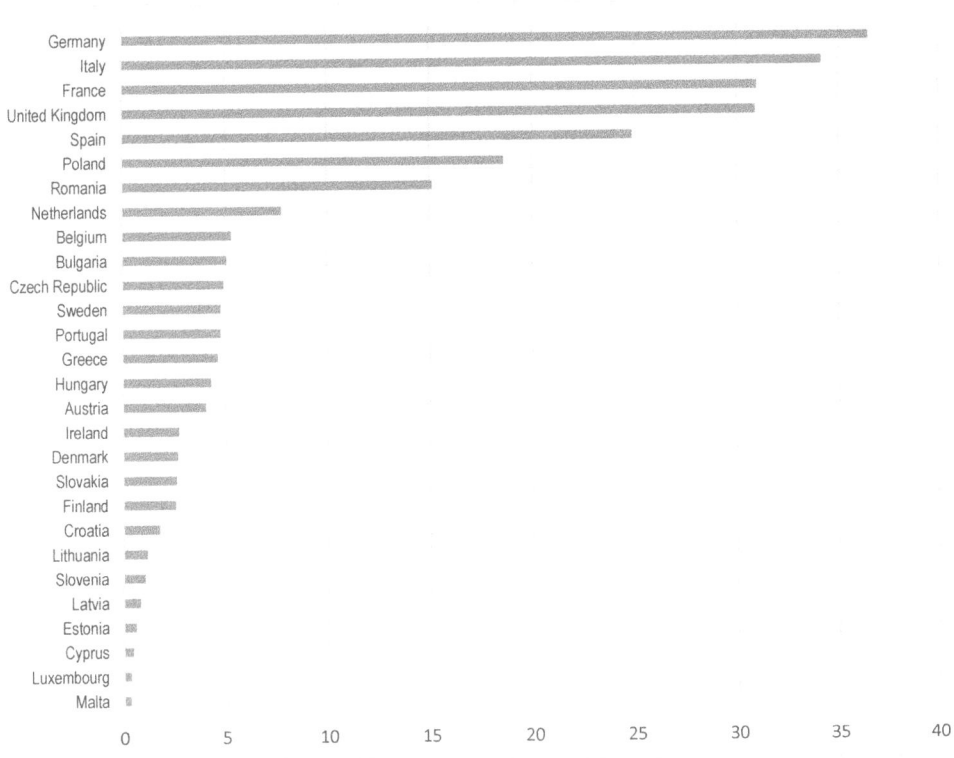

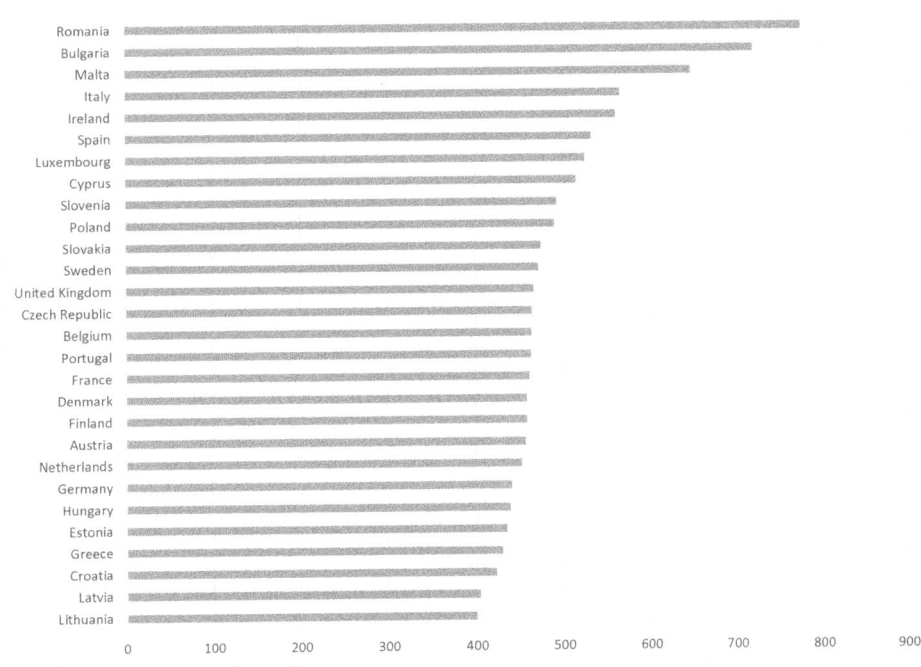

Source: OECD analysis based on European Commission (distance to compliance with UWWTD) and Eurostat (water-related public and household expenditures).

3.1.4. Summing up

The chart below brings together projections for water supply and for sanitation, combining the different scenarios: business as usual (driven by urbanisation), compliance with DWD and UWWTD, and efficiency (reduction of leakage in water supply). Aggregate figure for the 28 member states amount to EUR 289 billion.

Sanitation represents the lion's share of the total additional expenditures. This is particularly the case in Italy, Romania and Spain and - at lower levels – in Bulgaria, Croatia, Portugal and Slovakia. In these countries, urban population growth plays a minor part (sometimes nil) in projected future expenditures for water supply and sanitation.

Figure 3.5. Total cumulative additional expenditures by 2030 for water supply and sanitation

2020-2030, BAU + Compliance + efficiency (EUR billion)

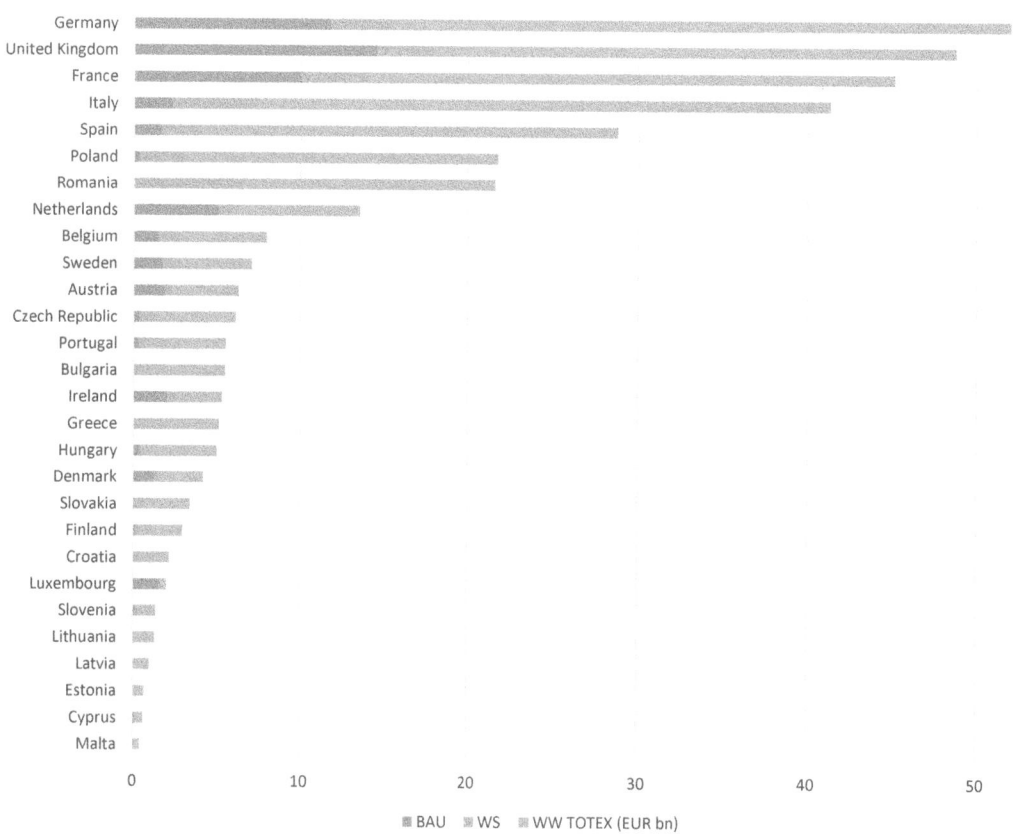

Source: OECD analysis based on European Commission and Eurostat data.

The picture is different when the size of the population is factored in. The Figure below projects per capita levels of expenditures for the same scenarios. For reasons already explained, Luxemburg stands out. In Ireland and Romania, inhabitants are projected to spend more than EUR 1,000 in addition to current levels of expenditures, between now and 2030. In most other countries, the additional level of expenditure per capita ranges between EUR 500 and 1,000. At the low end of the spectrum, projections for Greece, Hungary and Latvia may reflect optimistic reliance on IAS and an underestimate of additional expenditures required to comply with UWWTD. The situation in Lithuania may reflect significant un-used capacities for sanitation that result from high level of investment in recent decades and users' reluctance to connect.

Figure 3.6. Per capita cumulative additional expenditures by 2030

BAU + Compliance + efficiency (EUR)

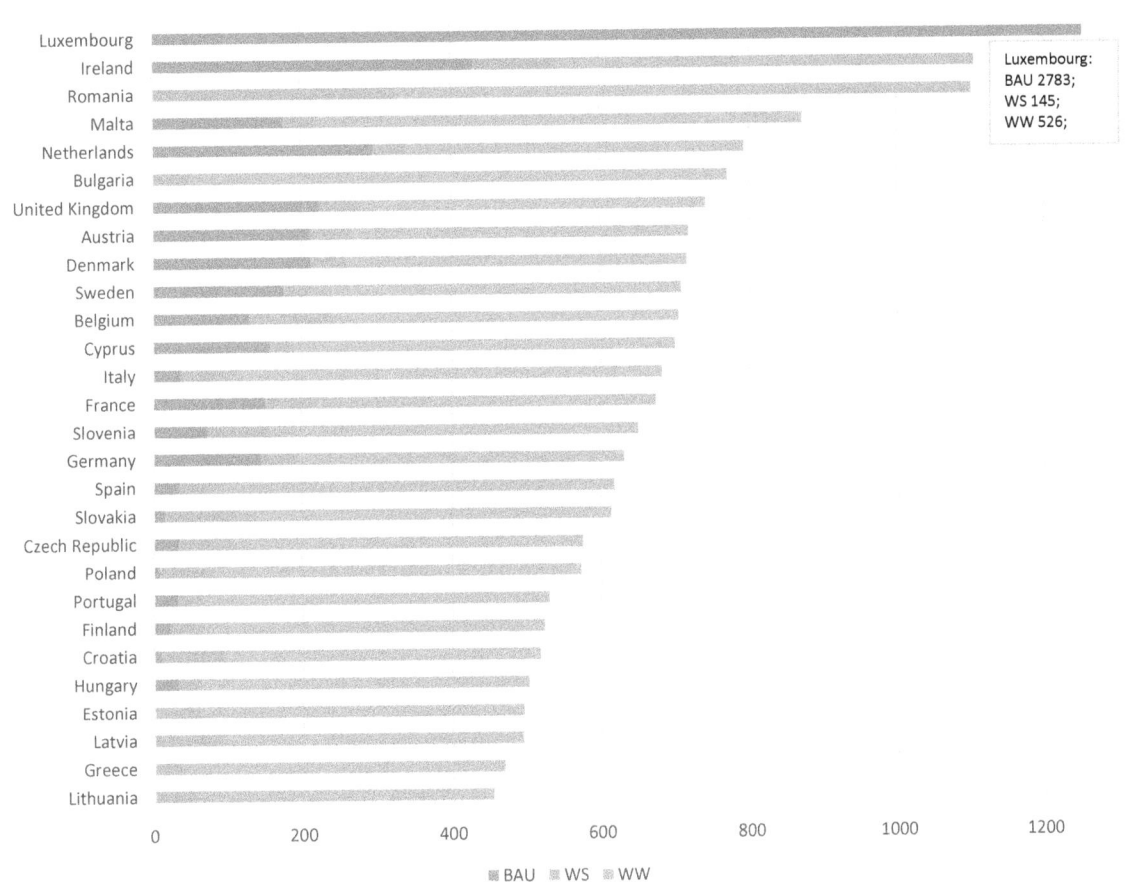

Source: OECD analysis based on European Commission and Eurostat data.

A telling indicator of the additional level of effort required by each country is to compare the additional expenditures for water supply and sanitation with the current level of expenditures as captured by the baseline. The chart below does so, on an annual basis. It is assumed that each country spreads the additional expenditures evenly over the period.

According to the projections, all countries (but Germany) will need to increase annual expenditures for water supply and sanitation by more than 25%. At the higher end, Romania and Bulgaria need to double (or more) the current level of expenditures. Finland is projected to increase expenditures by 85% (this may reflect the fact that the current level of expenditures in Finland is probably underestimated; see previous comment). At the lower end of the spectrum, Cyprus, the Czech Republic, France, Germany, the Netherlands, Slovenia are projected to face comparatively minor needs for increase (by less than 1/3). This is likely to reflect different situations, including high levels of expenditures and good anticipation of future needs, significant catch up in the recent decades (Czech Republic), or underestimate of future needs, possibly driven by overreliance on IAS (Slovenia).

Figure 3.7. Per Annum additional expenditures by 2030

BAU + Compliance + Efficiency vs. baseline

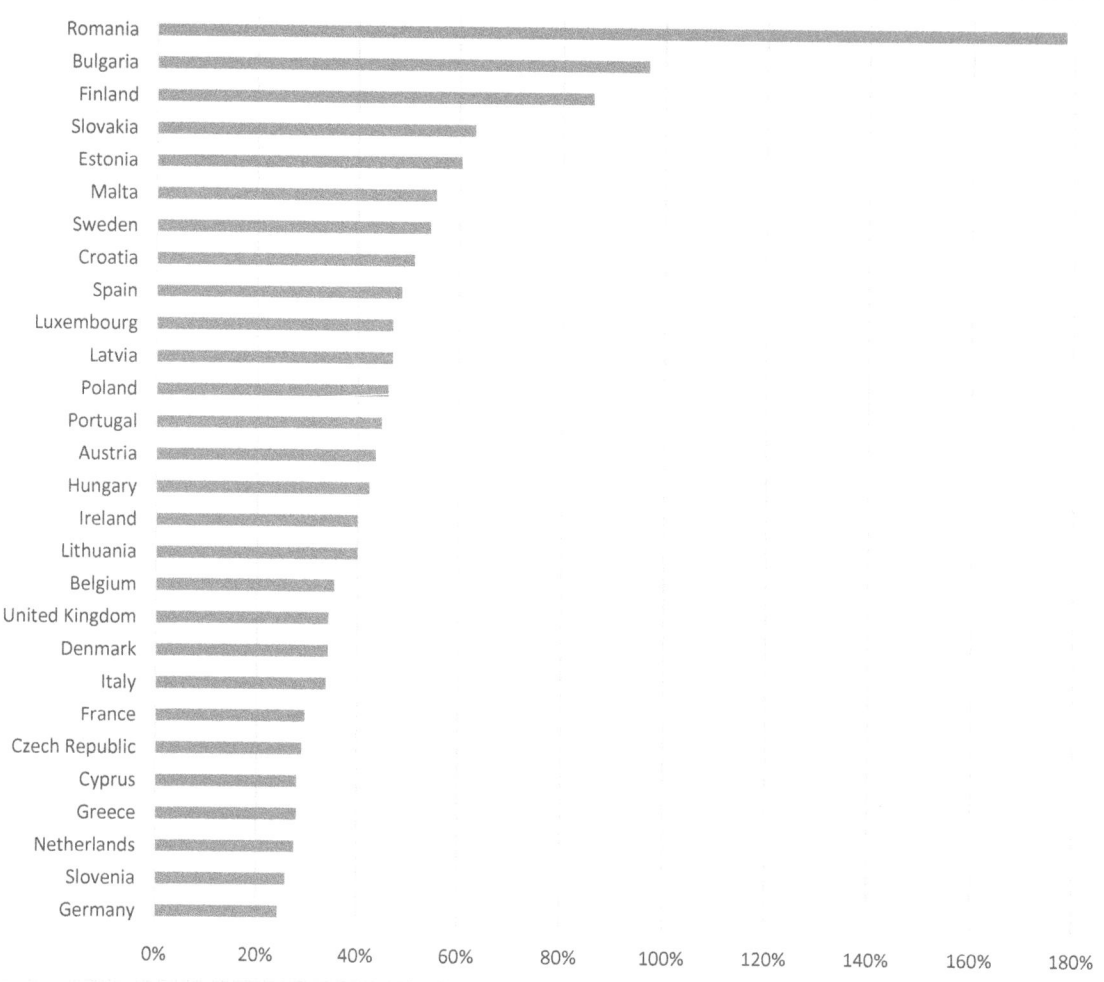

The subsequent part of the report will discuss how feasible such additional levels of effort are, considering the financing capacities of countries and room for manoeuvre.

3.2. Flood protection

This section covers two segments of flood protection:

- Riverine floods. As mentioned above, projections reflect the respective impact of climate change and of socio-economic factors, namely economic and demographic growth. These impacts are projected on three variables: the value of assets at risk of flooding, the number of people affected by floods, and the value of GDP affected by floods.
- Coastal floods. Coastal floods are captured qualitatively; quantified projections may be developed at a later stage, when a consistent set of data is released by the World Resources Institute.
- Urban floods are partially captured in the qualitative discussion of emerging challenges (Section 4.2).

As mentioned above, the inability to monetise investment needs results from the paucity of data on current level of expenditures for flood protection. It reflects two assumptions:

- The appropriate level of security against flood risk will remain stable over the period. This is a strong assumption, as the public opinion may be less willing to accept risks of floods as countries develop and people become more aware of what is at stake;
- The cost of mitigating flood risks rises at the same rate as the share of the population, the value of assets or GDP exposed to floods. This again is a strong assumption. As experience accumulates, countries may favour technologies and flood management techniques and policies which can become significantly more costly (large dykes) or alternatively invest in technics and policies that are comparatively less expensive or can generate multiple benefits (nature-based solutions). Alternative ways of mitigating floods risks are discussed in the final part of the report.

3.2.1. Projecting additional expenditures for protection against riverine flood risk

Figure 3.8 below shows the total growth factors for the three selected river flood risk indicators. A growth factor is defined as the factor by which current flood risk expenditures should be multiplied in order to maintain current flood risk protection standards in the future (by 2030).

Figure 3.8. Total growth factors for river flood risk expenditure by 2030

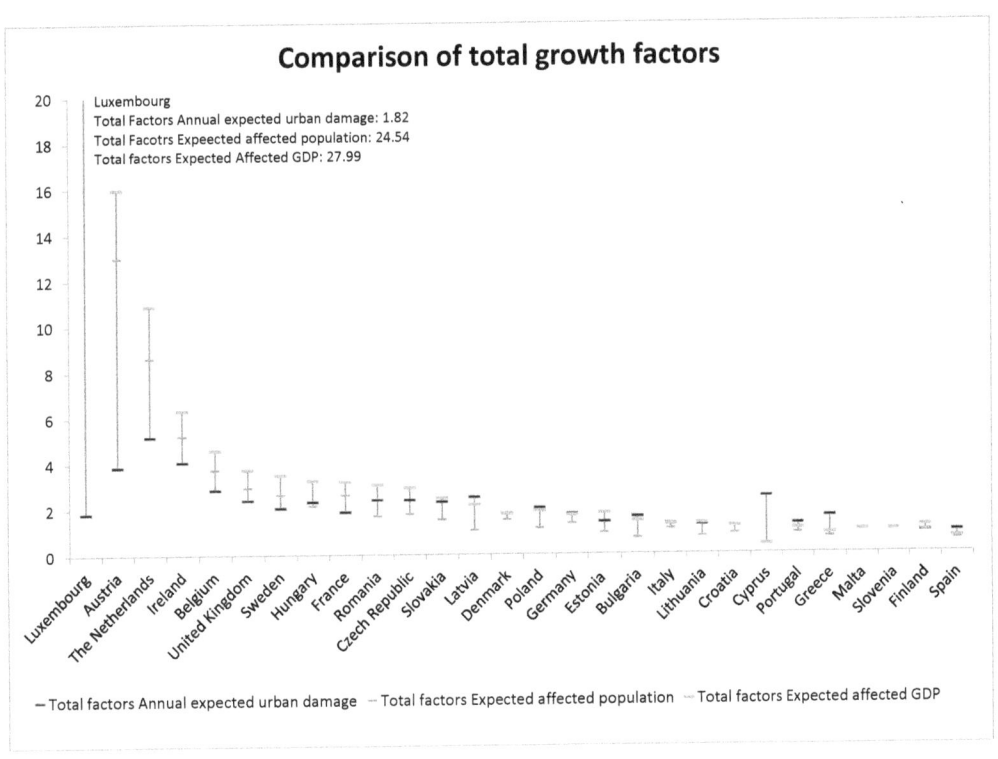

Source: Acteon, for this project, based on WRI projections.

Countries can be clustered in different categories, reflecting different perspectives on future exposure to riverine floods.

Table 3.2. Country clusters based on projected exposure to riverine floods

Countries affected by the highest total growth factors	Countries affected by moderate growth factors	Countries benefitting from lower exposure of population	Countries benefitting from low or negative growth factors
Austria, Luxembourg, the Netherlands	Belgium, Czech Republic, Denmark, France, Germany, Hungary, Ireland, Poland, Romania, Slovakia, Sweden, the UK	Bulgaria, Croatia, Estonia, Latvia, Lithuania	Cyprus, Greece, Malta, Portugal, Spain

Countries affected by the highest total growth factors

The results show that the total growth factors for Austria, Luxembourg, and the Netherlands are the highest compared to other member states. These countries will face the highest expenditures for flood protection by 2030, if they aim to maintain current flood protection standards. The increase in total growth factors is driven by climate change, indicating that urban assets, GDP and population will be increasingly exposed to flooding in the future compared to the current situation.

Countries affected by moderate growth factors

A large group of countries faces moderate growth factors – positive but lower than the growth factors for Austria, Luxembourg and the Netherlands: Belgium, Czech Republic, Denmark, France, Germany, Hungary, Ireland, Poland, Romania, Slovakia, Sweden and the UK. These countries will face increasing flood protection expenditures by 2030, if they aim to maintain current flood protection standards.

Climate change will significantly increase future flood risk even though the effect is less pronounced for some countries. For example, for Czech Republic, Denmark, Germany, Hungary, Poland, Romania and Slovakia, the impact of climate change is relatively low and more or less equal to the contribution of socio-economic developments in the explanation of future increases in flood risk.

These countries are exposed to a certain level of river flood risk due to their geographical characteristics, but are less vulnerable than Austria, Luxembourg and the Netherlands. Some economic activity and part of the population are located in floodplains that will face more frequent and severe flooding due to increased precipitation in winter. In the future, flood risk is expected to increase due to economic developments, population growth and urbanization in flood plain areas.

Countries benefitting from lower exposure of population

In Bulgaria, Croatia, Estonia, Latvia and Lithuania, the total growth factors for annual expected urban damage and annual expected exposed GDP are positive whereas the growth factor for expected affected population is negative. The countries have a level of economic development that is below the average of EU member states, but strong economic growth is expected in the future. Finally, the population is expected to decrease in many of these countries.

These countries will face slightly increasing flood risk expenditures by 2030, if they aim to maintain current flood protection standards. In contrast with other member states, socio-economic developments – not changes in the climate - have a relatively large contribution to a future increase in flood risk in these countries. This group of countries is exposed to river flood risk due to their geographical characteristics.

Countries benefitting from low or negative growth factors

Finally, in Cyprus, Greece, Malta, Portugal and Spain, several total growth factors are low or negative. Climate change is the dominant growth factor explaining the negative total growth factors for some indicators. In general, these countries have limited exposure to river flood risk due to their arid or semi-arid climate (even though some catchments are exposed to flooding during winter). Future flood risk is expected to decrease due to climate change and this is reflected in the negative growth factors for several indicators. These countries are projected to face no increase in flood risk expenditures by 2030.

3.2.2. Projecting additional expenditures for protection against coastal flood risk

The results of the qualitative assessment of projected coastal flood risk investment needs based on three vulnerability indicators (change of build-up in flood prone areas, number of people exposed to flooding, damage costs) are presented in Countries with high actual coastal flood risk will need to invest significantly in the future in order to maintain current flood protection standards compared to other member states. Several countries are actually exposed to high coastal flood risk: France, the Netherlands and the UK. These countries share the North Sea or the Atlantic Ocean Scenario, both maritime basins for which projections show that sea level rise is expected to be significant.

Climate change appears to be the dominant factor for an increase in future projected investment needs. Countries that have a high damage potential due to urban development, population and economic activity in the coastal flood plain have a higher vulnerability to flood risk. Due to socio-economic developments the flood risk in coastal flood plains could increase. However, the effect of socio-economic developments in explaining future coastal flood risk appears to be subordinate to the effect of climate change.

Table 3.3 below. A more comprehensive set of country-specific data that affect exposure to coastal floods is appended. Based on the evaluation of the three vulnerability indicators, countries were classified in one of four categories of projected coastal flood risk investment needs, in which 1 indicates very low growth of projected investment needs and 4 very high growth of projected investment needs by 2030.

Countries with high actual coastal flood risk will need to invest significantly in the future in order to maintain current flood protection standards compared to other member states. Several countries are actually exposed to high coastal flood risk: France, the Netherlands and the UK. These countries share the North Sea or the Atlantic Ocean Scenario, both maritime basins for which projections show that sea level rise is expected to be significant.

Climate change appears to be the dominant factor for an increase in future projected investment needs. Countries that have a high damage potential due to urban development, population and economic activity in the coastal flood plain have a higher vulnerability to flood risk. Due to socio-economic developments the flood risk in coastal flood plains could increase. However, the effect of socio-economic developments in explaining future coastal flood risk appears to be subordinate to the effect of climate change.

Table 3.3. Projected coastal flood risk investment needs

	Change in built-up in areas vulnerable to coastal floods	People in the 100-year flood plain	People flooded	Damage costs	Expenditures to protect against coastal flood risk
	%-increase	Million	Thousands/year	Billion Euro/year	Category 1-4
	2050	2030	2050	2050	
	Brown et al. (2011)	Neumann et al, (2015)	Hinkel et al, (2010)	Hinkel et al, (2010)	
Austria	-	-	-	-	-
Belgium	10,34	-	1,9	1,1	3
Bulgaria	0	-	0,2	<0,1	1
Croatia	-	-	-	-	-
Cyprus	60	-	0,1	<0,1	1
Czech Republic	-	-	-	-	-
Denmark	18,69	-	0,5	0,5	2
Estonia	0	-	0,1	<0,1	1
Finland	2,17	-	0,3	0,2	1
France	7,13	-	3,5	2,5	4
Germany	1,36	3,2	2	3	3
Greece	3,57	-	0,5	<0,1	1
Hungary	-	-	-	-	-
Ireland	21,43	-	0,6	<0,1	1
Italy	0	2,4	1,1	0,3	3
Latvia	0	-	0,8	<0,1	1
Lithuania	0	-	0,8	<0,1	1
Luxembourg	-	-	-	-	-
Malta	-	-	0,1	<0,1	1
Poland	25	-	4,5	<0,1	3
Portugal	4,55	-	0,7	0,2	2
Romania	0	-	1,1	<0,1	2
Slovakia	-	-	-	-	-
Slovenia	0	-	0,1	<0,1	1
Spain	3,64	1,6	1,6	0,4	2
Sweden	10,17	-	0,2		1
The Netherlands	8,54	10,2	5	2,3	4
United Kingdom	13,31	4,4	4,8	1,2	4

3.3. Investment needs under the Water Framework Directive

The DWD, UWWTD and Flood Directive are instrumental to compliance with the WFD. However, compliance with the three Directives covered in this report does not guarantee compliance with the WFD: more will need to be done to achieve good status. This section discusses qualitatively what will remain to be done, after countries comply with the three "technical" Directives.

It is worth noting that there may be some tensions across Directives, for instance when measures taken to mitigate flood risks affect environmental flows or the hydromorphology of rivers and lakes. Such situations were anticipated in the Water Framework Directive, which allows for exemption by application of the

Article 4.7, according to which deterioration of status or non-achievement of good status or potential can be justified under certain conditions.

3.3.1. EU Member States often fail to meet the water quality objectives of the WFD

Many EU countries fail to achieve 'good' chemical and ecological status of water bodies, as required under the EU Water Framework Directive (WFD)[1]. This is often despite compliance with technical EU water directives on drinking water, urban wastewater treatment and floods.

Based on the latest State of Water report by the European Environment Agency (2018):

- Only 40 % of the surface water bodies in the EU are in 'good' or 'high' ecological status. Lakes and coastal water bodies have a slightly better status (ca. 50%) than rivers and transitional water bodies (ca. 30-35%). The central European river basin districts, as well as some of the southern river basin districts, show the highest proportion of water bodies not achieving good ecological status or potential. The overall ecological status has not improved since the first reporting of River Basin Management Plans in 2009.
- Similarly, only 38 % of the surface water bodies in the EU are in 'good' chemical status. Almost half (46%) of the surface water bodies are not achieving good chemical status and 16% of the water bodies have unknown chemical status. High levels of mercury is a major cause of chemical status failure.
- Good chemical status of groundwater was achieved for 74 % of groundwater bodies.

Considering the large proportion of surface waters failing to meet 'good' ecological and chemical status, it is unlikely that the EU WFD objective of achieving good status of waters will be met by 2027 (when all exemptions have been used). Full implementation of the management measures under the WFD, in combination with full implementation of other relevant directives (e.g. Urban Wastewater Treatment, Nitrates Directive) is needed in order to restore the ecological and chemical status or potential of water bodies.

3.3.2. What prevents EU member states from achieving good water quality status

Failure to achieve good ecological and chemical status under the WFD primarily derives from three main pressures:

- *Diffuse (non-point) source pollution from rural and urban sources* (compliance with the Urban Wastewater Treatment Directive largely mitigates point source pollution). Diffuse source pollution affects the water quality of 62 % of surface water bodies and 41 % of groundwater bodies in the EU (EEA, 2018). Agricultural production is a major source of diffuse pollution. In Europe, diffuse source pollution is mostly due to excessive emissions of nutrients (nitrogen and phosphorus) and chemicals, such as pesticides. Atmospheric deposition is the leading source of mercury pollution[2] in most of the surface water bodies failing to achieve good chemical status. The EEA estimates that measures taken under the Nitrates Directive are not enough to tackle significant pressures from diffuse sources to reach good ecological status (EEA, 2018).
- *Alteration to the natural hydromorphology*[3] of rivers and lakes. Hydromorphological pressures are the second most commonly occurring pressures on surface water bodies (after diffuse source pollution) affecting 40% of all surface water bodies in the EU. In addition, 17 % of European water bodies have been designated as heavily modified (13%) or artificial water bodies (4 %) (EEA, forthcoming). Changes in the natural geomorphology and water flow of water bodies (e.g. channelised rivers disconnected from their floodplains, dams, canals, flood defences, reclaimed land) can have severe impacts on water quality, aquatic health, and the ability of ecosystems to

process and retain pollutants (EEA, 2018; Nilsson and Malm Renöfält, 2008; Wagenschein and Rode, 2008). For example, a study on the Weisse Elster River, Germany, revealed that the nitrogen retention rate is almost 2.4 times higher in a natural section of the river compared with a heavily modified and channelised section (Wagenschein and Rode, 2008).

- *Historical pollution, particularly* contaminated *river and lake bed sediment.* Historical pollution from industry and mining can often be a long-lasting source of pollution. Such pollution may have occurred when the science on the human and ecological health impacts was not clear, and when pollution regulations, monitoring and compliance were not as stringent as today. Historical pollution can be costly to remediate, as demonstrated in the case study of Flix, Ebro Basin, Spain (Box 3.2). One complication of historical pollution is the polluter is often no longer around to pay for the remediation of pollution, and thus the cost of clean-up is frequently left to governments and the tax payer.

These pressures affect the good functioning of water-related ecosystems, contribute to freshwater biodiversity loss, and threaten the long-term delivery of ecosystem services and benefits to society and the economy (e.g. the value of clean water and recreation).

Box 3.4. A case of costly historical pollution: Flix Reservoir, Ebro Basin, Spain

Accumulated historical contamination from industrial sources remains a persistent pollution source in the Ebro Basin, Spain - the second largest river flowing into the NW Mediterranean. This is nowhere more evident than in Flix Reservoir, which has been affected by toxic wastewater discharge from a chlor-alkali electrochemical plant since its establishment in 1897. As a result, elevated levels of organic contaminants, heavy metals and radioisotopes have accumulated in the water and sediment of the reservoir.

There was concern about the risk of flood, dam failure and the suspension of the contaminated sediment for transmission downstream - a potential threat to the water supplies of municipalities downstream, as well as the nearby protected Sebes natural reserve and the Ebro coastal delta system. Two possible options were considered:

1. Confine the contaminated sediment in the Flix reservoir (cheap)
2. Remove the sediment and treat the water from the reservoir, by changing the river flow and building a retaining/confinement wall.

A decision to extract, treat and eliminate the contaminated sludge and subsequently restore the Ebro River and its ecosystem was made in 2009 (option 2). Up to one million cubic meters of contaminated reservoir sediment will be removed by the project to reverse more than a century of pollution. The total cost is estimated at EUR 200 million - the largest investment ever for a decontamination project in Spain. It is estimated that the clean-up will take two years and eight months to complete.

The Flix Reservoir decontamination project draws 30% of its funds from the Spanish government and 70% from the European Union Cohesion Fund. There is also a Land Restitution Plan associated with the project, aimed at providing compensation for the people affected by the work. This plan entails EUR 57 million investment, split between the national government (36 million) and the Catalan government (21 million).

3.3.3. Options for investment to improve water quality

Mitigation and/or restoration measures are required to meet the WFD goal of achieving, enhancing or maintaining good status of transitional, coastal and freshwater bodies. Options for investment to improve water quality, including the targeted management of diffuse pollution, restoration of the natural hydromorphology and remediation of historical pollution, are outlined in Table 3.4.

Table 3.4. Examples of investment options to improve water quality

Investment type / Type of pollution targeted	Green (natural) infrastructure	Grey (built) infrastructure	Management practices	Policy and planning
Diffuse pollution	Wetlands Riparian planting Green roofs Permeable pavements Green swales Returning river systems to their natural state (river restoration) Afforestation of upstream catchments Land retirement (protected areas)	Phase out combined sewer overflows Build dual sewage and stormwater networks Construct stormwater storage systems (tunnels, reservoirs)	Cover crops Nutrient budgeting Fertiliser and pesticide efficiency Optimised manure management Responsible chemical storage Managing hotspots and at-risk vulnerable areas	Master plans or conservation plans for restoring the water quality and ecosystem health Ecological/ minimum flow requirements Removal of harmful subsidies Economic instruments (e.g. PES, pollution charges, water quality trading) Regulations (e.g. drinking water and wastewater standards, restrictions or bans on harmful chemicals, land use restrictions) Advisory services and knowledge-building
Alteration to natural hydromorphology	River restoration Afforestation of upstream catchments Land retirement (protected areas)	Removal of obstacles and structures Installation of fish passes or ladders Upgrade wastewater treatment plants	Riparian planting to stabilise river banks Managing hotspots and at-risk vulnerable areas	Master plans or conservation plans for restoring the water quality and ecosystem health Ecological/ minimum flow requirements
Historical pollution	Bioremediation (i.e. microbial biodegradation) Natural attenuation	Dredging to remove and treat contaminated sediments Water purification methods (e.g filtration, air stripping or thermal treatment)	Containment Managing hotspots and at-risk vulnerable areas	Master plans or conservation plans for restoring the water quality and ecosystem health Ecological/ minimum flow requirements Water safety plans Environmental Impact Assessments Site remediation plans Chemical spill response plans Liability or insurance requirements Pollution fines and penalties

The Water Framework Directive (Article 5) requires Member States to carry out an economic analysis to identify the most cost-effective responses to pressures on the status of water bodies. The second generation of river basin management plans indicates that progress in this direction has been slow.

There is a case for the utilisation of cost-effective prevention and abatement practices that could yield more beneficial results in terms of water quality improvements and control-cost savings (Shortle et al., 2012; Shortle and Horan, 2013). This is for instance the case with measures related to controlling diffuse pollution from agriculture, which have triggered mixed results (see OECD, 2017 for a more detailed discussion).

Innovative approaches, such as water quality trading and other economic instruments, offer the possibility of improving the effectiveness and efficiency of water quality programmes (see OECD, 2017 for more information and case studies). Figure 3.9 demonstrates the variation in cost of various management and infrastructure options to reduce nitrogen loading in Chesapeake Bay, United States.

Figure 3.9. Cost comparison of options to reduce a nitrogen loading, Chesapeake Bay Watershed, United States

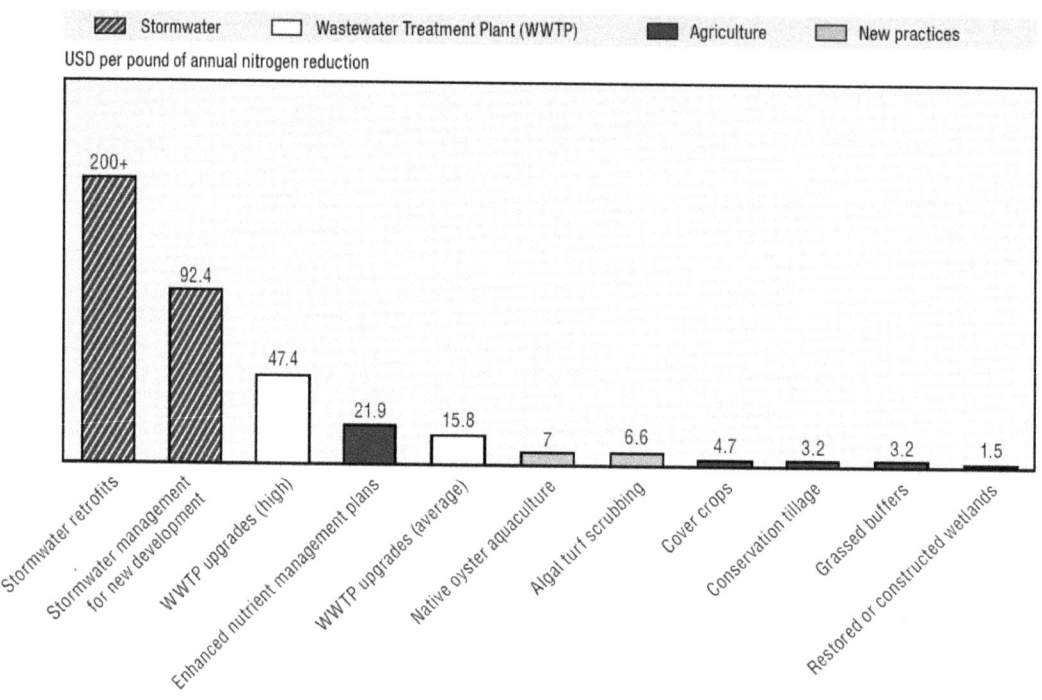

Source: Jones, C. (2010), How Nutrient Trading Could Help Restore the Chesapeake Bay, WRI Working Paper. World Resources Institute, Washington, DC.

The fundamental challenge for policy makers is to understand the – economic, social and environmental - costs of non-compliance, and to compare the costs of measures with the value they create for the communities. In the context of the Blue 2 study, Russi and Farmer (2018) test a methodology to assess the costs and benefits of the implementation of the EU water acquis in selected river basins.

A set of well-established principles can guide the design and implementation of policy responses to water pollution (OECD, 2017).

- The Principle of Pollution Prevention derives from the fact that prevention of pollution is often more cost effective than treatment/remediation options.
- Similarly, the Principle of Treatment at Source reflects the observation that the later the stage of control, the less effective and more costly the treatment is likely to be due to pollution dispersion.
- The Polluter Pays Principle makes pollution costly and incentivises reductions.
- The Beneficiary Pays Principle allows sharing of the financial burden of water quality management when necessary.
- Equity should be considered with regards to fair allocation of pollution rights, costs and benefits of abatement, and the needs of future generations.
- Policy coherence is required to ensure initiatives taken by different policy sectors (e.g. agriculture, urban planning, and climate) do not have negative impacts on water quality and to capitalise on co-benefits from water quality interventions. Investments in green (nature-based) infrastructure solutions have advantages here; they can support the goals of multiple policy areas, increase the resilience of ecosystems, and are generally less capital intensive and have lower operation,

maintenance and replacement costs than grey (built) infrastructure alternatives (see Sections 12.1, 12.4).

Box 3.5 presents a New Zealand case study illustrating how the strong endorsement of a voluntary agreement with the dairy industry helped to spur significant investment in the reduction of diffuse source water pollution and laid the groundwork for forthcoming national regulation on water quality (for the dairy industry and other sectors, e.g. beef cattle, sheep, deer, pigs). While the combination of tools is sophisticated, it illustrates considerations that contribute to achieving good ecological and chemical status of water bodies at the least costs for communities.

Box 3.5. A voluntary agreement to stimulate investment in the protection of water bodies, in New Zealand

In recognition of the need for limits on water quality and resource allocation, the New Zealand government issued the National Policy Statement for Freshwater Management (NPS-FM) in 2011 (subsequently amended in 2014). In 2014, the government announced its intention to require the exclusion of dairy cattle from waterways by 1 July 2017 (MfE, 2016). It was at this time, that the "Sustainable Dairying: Water Accord" was established as a voluntary agreement between government and the dairy industry (DairyNZ and DCanz, 2013). The accord sets clear environmental performance targets for fencing off dairy cattle from water bodies; the establishment of riparian areas; the management of nutrients, effluent and water use; and environmental measures for farm conversions to dairy.

Since the accord's inception, dairy cattle have been excluded from 97.2% of New Zealand's waterways that are subject to the accord[1]. Greater than 99% of 44 386 regular livestock river crossing points on dairy farms have bridges or culverts to protect local water quality, while 83% of dairy farms have nutrient budgets (DairyNZ and DCanz, 2017). Farmers have spent more than over NZD 1 billion (EUR 580 million) on environmental initiatives over the last five years, with the majority of investments (70%) on effluent system upgrades and fencing (DairyNZ and DCanz, 2016). In addition, NZD 10 million (EUR 5.8 million) has been spent on environmental stewardship and farmer support programmes covering research, development, and farmer extension (DairyNZ and DCanz, 2017).

Through the Accord, tangible results have been achieved ahead of the adoption of relevant government regulation, which require public consultation and are open to potentially lengthy court cases[2]. The Accord has also helped create acceptance before becoming regulation and has contributed to the design of the regulation for sectors beyond dairy (i.e. beef cattle, sheep, deer, pigs).

Notes:
1. The total number of farms covered by the Accord is approximately 11 400, representing 95% of all New Zealand dairy farms (DairyNZ and DCanz, 2017).
2. In 2016, the Government proposed a set of national regulations requiring exclusion of dairy cattle, beef cattle, deer and pigs from water bodies by dates ranging from 2017 (dairy and pigs) to 2030 (beef and deer on lowland/rolling hills (MfE, 2016).
Source: DairyNZ and DCanz, 2013; 2016; 2017, MfE, 2016.

References

Brown S., Nicholls R.J., Vafeidis A., Hinkel J., and Watkiss P. (2011), The Impacts and Economic Costs of Sea-Level Rise in Europe and the Costs and Benefits of Adaptation. Summary of Results from the EC RTD ClimateCost Project, in Watkiss P. (Editor), *The Climate Cost Project. Final Report. Volume 1: Europe*, Stockholm Environment Institute, Sweden, ISBN 978-91-86125-35-6.

DairyNZ and DCanz (2017), *Sustainable dairying – Water Accord: Three years on… Progress report for the 2015/16 season*.

DairyNZ and DCanz (2016), *Sustainable Dairying: Water Accord: Two Years on… Progress report for the 2014/15 season*.

DairyNZ and DCanz (2013), *Sustainable Dairying: Water Accord*, DairyNZ, Hamilton.

Eureau (2018), *Briefing Note. Update on the 3TS*, Brussels, Belgium.

European Commission (2017), *Study supporting the revision of the EU drinking water directive. Final impact assessment report. Part II, impact assessment*, Brussels

European Commission (2015), *Good Practices on Leakage Management. Main Report*, Brussels

European Court of Auditors (2018), *Special report no 25/2018: Floods Directive: progress in assessing risks, while planning and implementation need to improve*, Luxemburg.

European Environment Agency (2018), *State of Water report 2018*, European Environment Agency, Copenhagen.

European Investment Bank (2016), *Restoring European Competitiveness*, Luxemburg

Freyberg, T. (2013), Contaminated sludge clean-up begins on Spain's Ebro River to reverse toxic legacy https://www.waterworld.com/articles/2013/04/contaminated-sludge-cleanup-begins-on-spains-ebro-river-to-reverse-toxic-legacy.html [accessed 15/05/20184].

Grantham, T.E., Figueroa, R. and N. Prat (2012), Water management in Mediterranean river basins: A comparison of management frameworks, physical impacts, and ecological responses. Hydrobiologia, pp. 1-32.

Hinkel J., Nicholls R.J., Vafeidis A.T., Tol R.S.J., Avagianou T. (2010), *Assessing risk of and adaptation to sea-level rise in the European Union: An application of DIVA*. Mitig Adapt Strateg Glob Change 15:03-719

Jones, C. (2010), How Nutrient Trading Could Help Restore the Chesapeake Bay, WRI Working Paper. World Resources Institute, Washington, DC.

MfE (2016), *Next Steps for Fresh Water: Consultation document*, Ministry for the Environment, Wellington.

OECD (2017), *Diffuse Pollution, Degraded Waters: Emerging Policy Solutions*, OECD Studies on Water, OECD Publishing, Paris. http://dx.doi.org/10.1787/9789264269064-en

Neumann B., Vafeidis A., Zimmermann J., Nicholls R.J. (2015), *Future Coastal Population Growth and Exposure to Sea-Level Rise and Coastal Flooding - A Global Assessment*, DOI: 10.1371/journal.pone.0118571

Nilsson, C., and B. Malm Renöfält (2008), Linking flow regime and water quality in rivers: a challenge to adaptive catchment management, *Ecology and Society* Vol. 13, No. 2.

Palanquesa, A. et al. (2014), Massive accumulation of highly polluted sedimentary deposits by river damming, *Science of The Total Environment*, Vol. 497-498, p. 369-381.

Russi D., A. Farmer (2018), *Testing a methodology to assess the costs and benefits of the implementation of the EU water acquis in selected river basins*, Deliverable to Task A3 of the BLUE 2 project "Study on EU integrated policy assessment for the freshwater and marine environment, on the economic benefits of EU water policy and on the costs of its non- implementation". Report to DG

ENV.

Shortle, J.S. et al. (2012), Reforming agricultural nonpoint pollution policy in an increasingly budget-constrained environment, *Environmental Science and Technology*, Vol. 46, pp 1316-1325.

Shortle, J.S. and R.D. Horan (2013), Policy Instruments for Water Quality Protection, *Annual Review of Resource Economics*, Vol. 5, pp. 111-138.

Wagenschein, D. and M. Rode (2008), Modelling the impact of river morphology on nitrogen retention-A case study of the Weisse Elster River (Germany), *Ecological Modelling*, Vol. 211, pp. 224-232.

Notes

[1] The goal of the Water Framework Directive (WFD) is that good status should be achieved, enhanced or maintained in transitional, coastal and fresh waters. This goal is primarily concerned with the quality of surface and groundwater bodies: i) good ecological status in surface water bodies and ii) good chemical status in surface and groundwater bodies. The control of water quantity is also required to serve the objective of ensuring good quality (i.e. ensuring sufficient environmental flows for pollution dilution).

[2] Common atmospheric (diffuse) pollution sources of mercury include fossil fuel combustion (in particular coal-fired power plants), historic and current gold and silver mining operations, and natural sources (such as volcanoes, forest fires, and particulate and gaseous organic matter emissions from land and marine plants).

[3] Hydromorphology is a term used in river basin management to describe the interactions between hydrologic processes (water flow), geomorphic processes (landforms and earth materials) and the attributes of rivers, lakes, estuaries and coastal waters.

4 The capacity to finance projected investment needs across member states

This chapter discusses the capacities of EU member states to cover the financing needs presented in the previous chapter. The focus is on revenues from tariffs for water supply and sanitation services, public expenditure and commercial debt. The chapter concludes by ranking EU member states based on the severity of the financing challenge they face, as regards water supply, sanitation and, in different ways, flood protection. The subsequent chapter discusses a range of options to address the financing challenge.

This part of the report discusses the capacities of 28 member states to finance projected expenditure needs to comply with the revised Drinking Water, Urban Waste Water and Flood Directives. It discusses in succession affordability constraints that affect the capacity to raise additional through tariffs for water supply and sanitations services; capacities to raise additional public finance to cover water-related expenditure needs; and experience with - and opportunities for - mobilising commercial debt. The situation in each country is characterised and used to benchmark EU member states' relative capacity to finance projected expenditure needs to comply with the three Directives mentioned above.

The next part of the report explores options to minimise financing needs, make the best use of existing assets and financial resources, and harness additional sources of finance, as required.

4.1. Financing capacity

Projecting future volumes of available financing would be both a highly challenging and uncertain exercise. The approach taken to assess countries' financing capacities in this study was to consider three sets of complementary indicators – on the share of water bills in households' disposable income, public debt as a share of GDP, and domestic credit to the private sector as a share of GDP - thereby highlighting possible or likely latitude and constraints. Annex A provides further information about the data sources used for each indicator.

4.1.1. Raising tariffs for WSS services

With one exception in Europe, revenues from tariffs are considered a reliable source of finance to cover at least some of the costs of water supply and sanitation services. As shown in Part 2 of the report, the level of recovery of costs through revenues from tariffs vary. Affordability concerns are claimed to constrain a progressive move towards higher (full) recovery of costs of service provision through revenues from tariffs.

Figure 4.1 simulates the impact of passing on the additional expenditures for WSS to households (in this case for the disposable household income of the top-end of the lowest decile). This is based on current levels of expenditure for water supply and sanitation, augmented by projected additional level of effort (as a share of current level of expenditure; see Figure 3.6 above) and compared with the current households' disposable income. Obviously, this leads to an overestimation of affordability constraints, as households' disposable income is set to increase, driven by economic growth. However, projections of increase in income are too hazardous to be reflected in this discussion. Environmental and resource costs of service provision are not factored in. As mentioned above, data is not available for Croatia and Sweden. Ireland is in its own league as a vast majority of the population does not pay a water bill.

The Figure suggests that the 10% poorest households in 5 or 6 countries face affordability issues. But half of EU member states would face affordability issues for at least 5% of the population. This shows different levels of vulnerability to tariff increases across countries, and affects the way accompanying measures should be designed to mitigate the social consequences of higher prices.

Figure 4.1. Projected affordability issues – constant households disposable income

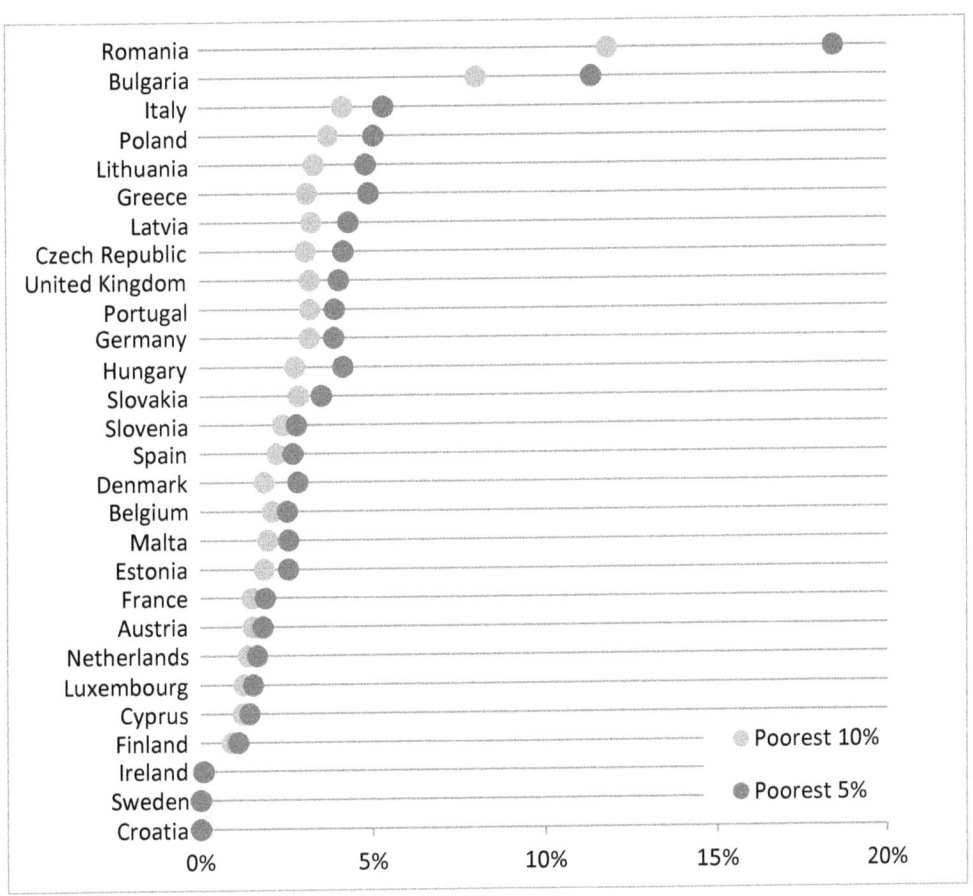

Note: Lack of household expenditure data for Croatia and Sweden. Known underestimate of total expenditures for Finland and Sweden. Households' disposable income is constant at 2011-15 level.
Source: EUROSTAT household expenditures and income data (2011-2015).

To complement a solely price-based analysis of affordability, Figure 4.2 adds a variable reflecting the percentage of the population at-risk of poverty. The top right corner of the scatter plot is obviously the least desirable position to be in as there is a higher risk in these countries that a significant share of the population becomes poor, thus raising more concerns about affordability.

Figure 4.2. Affordability of water supply and sanitation compounded by risk of poverty

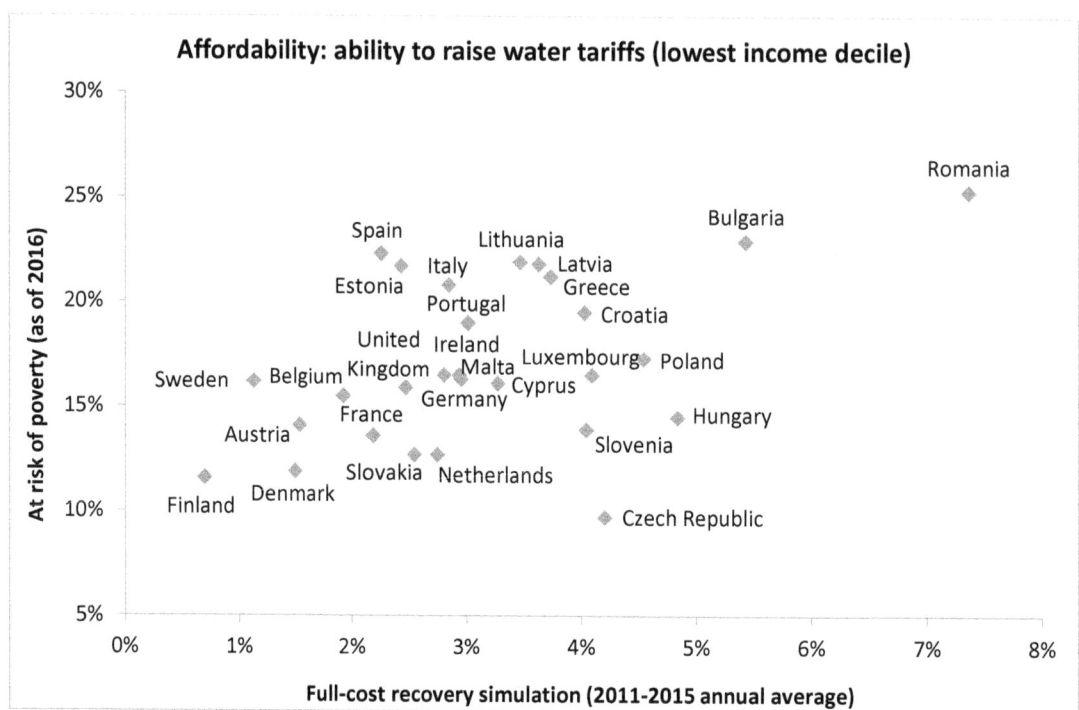

Note: Known underestimate of total expenditures for Finland and Sweden.
Source: Joint Research Centre, European Commission for prices (based on latest available year), EUROSTAT for household expenditures (2011-2015).

4.1.2. Increasing public spending

It is assumed that public spending for WSS may be increased based on either taxes or borrowing. Figure 4.3, therefore, combines countries' current level of taxation and indebtedness (both consolidated across various levels of public authorities and expressed as percentage of GDP).

A high percentage of taxes in GDP may both highlight a country's demonstrated ability to use taxation as an instrument to finance public expenditures as well as indicate a constraint to further increase taxes moving forward (and conversely for countries with a currently low percentage). Depending on its level, a high percentage of public debt may indicate a possible or likely budgetary constraint, which could prevent the country from increasing public spending and from borrowing at a reasonable cost.

Figure 4.3. Ability to increase public spending based on raising taxes or borrowing

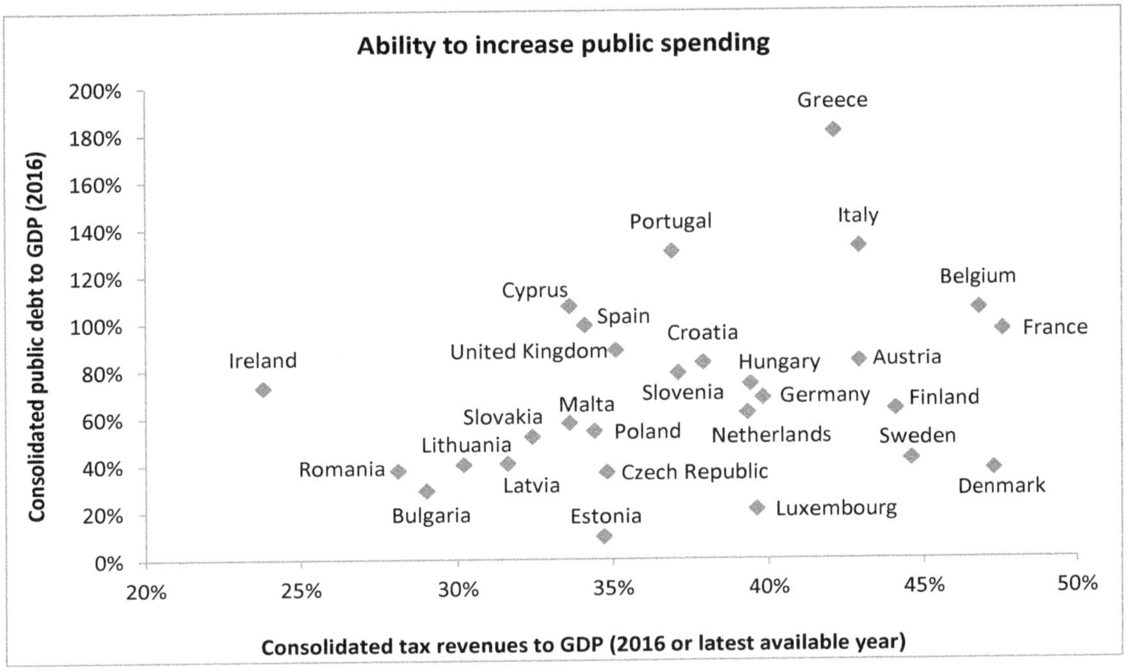

Ability to increase public spending

[Scatter plot: x-axis "Consolidated tax revenues to GDP (2016 or latest available year)" from 20% to 50%; y-axis "Consolidated public debt to GDP (2016)" from 0% to 200%. Country positions approximately:
- Greece: ~42%, 185%
- Portugal: ~37%, 133%
- Italy: ~43%, 135%
- Belgium: ~47%, 110%
- France: ~48%, 100%
- Cyprus: ~33%, 108%
- Spain: ~34%, 100%
- Croatia: ~38%, 85%
- Hungary: ~39%, 78%
- Austria: ~43%, 85%
- Ireland: ~24%, 73%
- United Kingdom: ~34%, 90%
- Slovenia: ~37%, 75%
- Germany: ~40%, 70%
- Finland: ~44%, 68%
- Slovakia: ~32%, 53%
- Malta: ~34%, 60%
- Poland: ~37%, 55%
- Netherlands: ~39%, 60%
- Sweden: ~45%, 45%
- Lithuania: ~30%, 42%
- Romania: ~28%, 38%
- Latvia: ~32%, 38%
- Czech Republic: ~35%, 38%
- Luxembourg: ~40%, 22%
- Denmark: ~48%, 38%
- Bulgaria: ~29%, 30%
- Estonia: ~35%, 13%]

Source: EUROSTAT (2016).

Given the heavy reliance of most member states on borrowing to finance part of their overall expenditures, Table 4.1 puts in perspective the consolidated public debt indicator displayed in Figure 4.3 above, by listing member states' current sovereign credit rating. In essence, the sovereign rating indicator illustrates whether a country is in a position to easily borrow and to do so at a reasonable cost. While the two indicators are coherent for quite many member states, it can be observed that some countries with relatively higher level of indebtedness are nevertheless assessed as slightly more risky (and vice versa).

Table 4.1. Sovereign credit rating

Rating	Description	Member States
AAA	Highest rating assigned by S&P. The obligor's capacity to meet its financial commitments on the obligation is extremely strong.	Denmark, Germany, Luxembourg, Netherlands, Sweden
AA	Only differs from the highest-rated obligations to a small degree. The obligor's capacity to meet its financial commitments on the obligation is very strong.	Austria, Belgium, Czech Republic, Estonia, Finland, France, Slovenia, United Kingdom
A	Somewhat more susceptible to the adverse effects of changes in circumstances and economic conditions than obligations in higher-rated categories. The obligor's capacity to meet its financial commitments on the obligation remains strong.	Ireland, Latvia, Lithuania, Malta, Poland, Slovakia, Spain
BBB	Adequate protection parameters. However, adverse economic conditions or changing circumstances are more likely to weaken the obligor's capacity to meet its financial commitments.	Bulgaria, Croatia, Cyprus, Hungary, Italy, Portugal, Romania
BB	Faces major ongoing uncertainties or exposure to adverse business, financial, or economic conditions that could lead to the obligor's inadequate capacity to meet its financial commitments.	
B	More vulnerable to non-payment than 'BB', but the obligor currently has the capacity to meet its financial commitments. Adverse business, financial, or economic conditions will likely impair the obligor's capacity or willingness to meet its financial commitments.	Greece

Source: Standards and Poors (2019).

In most EU member states, sub-sovereign entities and local authorities play an important role in covering financing needs in water supply, sanitation and flood protection. Available data sources do not allow monitoring their level of spending for flood protection. And they do not allow a robust assessment of room for manoeuvre to mobilise additional finance from these public authorities.

4.1.3. Tapping into private finance

As already illustrated in Part II some member states have been partly relying on debt financing to finance (mainly) upfront capital investments. In order to provide a more general indication of each country's ability to tap into domestic commercial debt financing, Figure 4.4 presents domestic credit to private sector as a percentage of GDP, compiled by the World Bank. This refers to financial resources provided to the private sector by financial corporations, such as through loans, purchases of non-equity securities, and trade credits and other accounts receivable, that establish a claim for repayment. A relatively high percentage may indicate that commercial debt is readily available in the country for financially sustainable WSS projects (and vice versa).

Figure 4.4. Domestic credit to private sector

(as % of GDP, 2015)

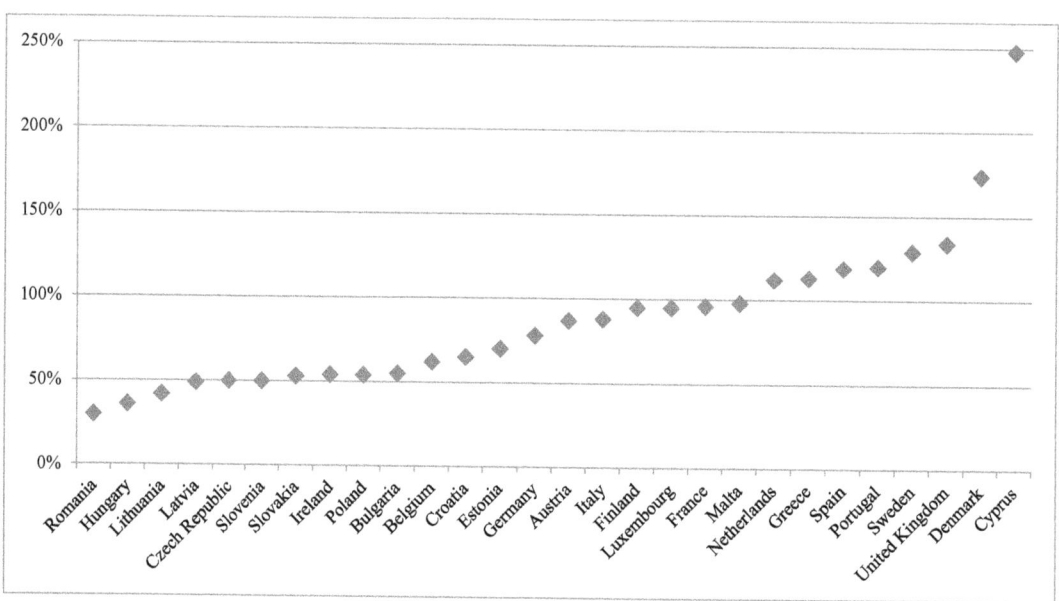

Source: World Bank's Ease of Doing Business Index database (2017), World Bank's Global Financial Development Database (2017).

The capacity to access commercial debt financing is unevenly spread among project owners within countries. For instance, while large utilities and operators typically have access to debt financing, a potential barrier relates to the fragmented nature of local authorities and the small size of projects. Further, the domestic credit to private sector as a percentage of GDP may not be an appropriate indicator e.g. for countries with disproportionate banking sectors compared to the size of the economy or with debt-financed real estate "bubbles" (e.g. Cyprus).

4.2. Preliminary conclusions

Table 4.2 below characterises the challenges member states face to finance projected investment needs for water supply and sanitation. Because expenditure needs to protect against flood risk could not be monetised, they are not considered in this Table.

Countries are ranked according to the additional level of effort required, compared to the baseline. The share of the baseline in GDP captures the current level of effort.

- Topping the list, Romania and Bulgaria face severe financing challenges as the projected additional level of effort is very high and rooms for manoeuvre for financing appear limited.
- Finland ranks high in terms of additional level of effort compared to current level of expenditures, but i) this reflects low level of effort in recent years and ii) Finland has room for manoeuvre to cover these additional expenditures.
- Slovakia and Estonia may face similar levels of effort in the future but Estonia is better placed to cover them, as public finance look less strained, should they need to be mobilised.
- Latvia, Poland and Portugal face similar levels of efforts in the future, but have distinct capacities to cover them. Affordability issues are relatively less severe in Portugal and access to private debt is easier.
- The ranking of Greece and Slovenia begs questions. The additional level of effort is probably underestimated, reflecting excessive reliance non IAS. A reassessment of additional financing needs would translate into severe challenges, as financing capacities are limited for both countries.
- The Netherlands and Germany are in privileged situations, as the additional level of efforts is comparatively limited and financing capacities are strong.

Previous parts of the report have signalled caveats and data limitations. While more detailed and accurate data may be available for any given country, the indicators and proxies used in the report are the best available to compare robust data across 28 EU member states, which is the level of ambition of the Table below. The Table has been used to identify countries which face the most severe challenges to finance projected expenditure needs to comply with the DWD, UWWTD and Floods Directive. Bulgaria, Croatia, Cyprus, Lithuania, Poland, Romania, Slovakia, Slovenia, and Spain were selected on that basis. The OECD and the European Commission then convened dedicated workshops in each of the selected countries, to fine-tune the understanding of the challenge and explore policy options to address it. Part V below captures the main outcomes of that stage of the project.

Table 4.2. Member states' capacity to finance projected investment needs for WSS

Country	% increased investment needs	% current expeditures in GDP	Raising tariffs	Raising public spending	Accessing domestic debt
Romania	178%	0.80%			
Bulgaria	97%	1.10%			
Finland	86%	0.20%			
Slovakia	63%	0.70%			
Estonia	60%	0.50%			
Malta	55%	1.00%			
Sweden	54%	0.30%			
Croatia	51%	0.90%			
Spain	49%	0.50%			
Luxembourg	47%	0.80%			
Latvia	47%	0.70%			
Poland	46%	1.10%			
Portugal	45%	0.70%			
Austria	44%	0.50%			
Hungary	42%	1.10%			
Ireland	40%	0.70%			
Lithuania	40%	0.60%			
Belgium	36%	0.60%			
Italy	34%	0.70%			
United Kingdom	34%	0.70%			
Denmark	34%	0.50%			
France	30%	0.80%			
Czech Republic	29%	1.30%			
Cyprus	28%	1.20%			
Netherlands	28%	0.90%			
Greece	28%	0.80%			
Slovenia	26%	1.40%			
Germany	24%	0.80%			

Source: Authors' calculation and interpretation.

The situation is more complex when financial needs to protect against flood risks are considered. The following clusters derive from the analyses of projected flood-related risks above:

- Member states listed in Category 4: France, the Netherland, the UK.
- Member states listed in Category 3: Belgium, Germany, Italy, Poland.

None of the countries listed in Category 4 stand out as facing particularly dire challenges related to financing future investment needs for water supply and sanitation. Poland and, to a lesser extent, Italy feature in Category 3 and at the same time are projected to face difficulties meeting financing requirements for water supply and sanitation.

Such a crude assessment of countries' capacities to cover financing needs for water supply, sanitation and flood protection is considered a basis for discussion only. More fine-grained analyses in selected countries can help to check whether projections reflect the actual situation. Preliminary discussions signal that the current level of effort does not reflect a potential backlog for investment. Assessment of distance to compliance or efficiency of water supply services only partially capture the performance of existing assets and the need to further invest. Anecdotal evidence suggests that in selected countries, ageing networks are likely to be the single biggest driver for investment in water supply and sanitation (see WAREG, 2017).

The appropriate level of effort can only be known with accuracy when member states compile robust knowledge on the state of the assets.

More fine-grained analyses can also support exploration of options to minimise financing needs and harness additional sources of finance. The following part of the report discusses some of them.

Such options can reflect how much water contributes to - and benefits from - broader economic development. They can also reflect how investments in other sectors contribute to water; this is potentially the case for land use and urban development; energy supply and climate change mitigation; or adaptation to climate change.

On-going work under the umbrella of the Roundtable on Financing Water can support such analyses. In particular, the way water is valued in society and the economy can drive investment decisions and willingness-to-pay of stakeholders who benefit from improved access to water supply and sanitation, flood protection, and more generally good ecological status of water bodies.

References

WAREG (2017), An Analysis of Water Efficiency KPIs in WAREG Member Countries, WAREG

5 Selected options to address financing challenges

This chapter explores options EU member states may wish to consider to minimise financing needs related to water supply, sanitation and flood protection, make the best use of available assets and financing resources, and harness new sources of finance, if and when required.

The options were discussed in a series of country workshops co-convened by the OECD and the European Commission in countries, which face the most severe financing challenges. Some of the options reflect the work of the Roundtable on Financing Water, a joint initiative by the OECD, the Netherlands, the World Water Council and the World Bank (more information on the Roundtable is available here).

Part III of the report projected expenditure needs EU members face to comply with the Drinking Water, Urban Wastewater Treatment and Flood Directives. While all countries need to increase the current level of expenditure for water supply and sanitation by 20% or more (see Figure 3.6), some face more challenging needs, including Finland (+85%)[1], Bulgaria (+100%) and Romania (+180%). It is likely that the investments remaining to be made for most countries to reach compliance are also among the most difficult: it is reasonable to assume that low hanging fruits have already been picked and future investments need to deal with the new connections with high marginal, or municipalities with low capacity.

The situation is compounded by projected needs to finance flood protection. With the exception of few arid Mediterranean countries (Cyprus, Greece, Malta, Portugal and Spain), other EU member states will need to increase flood protection expenditures if they want to keep pace with rising exposure to flood risks in the coming decades[2] (see Figure 3.7). Austria, Luxembourg and the Netherlands are exposed to the highest increase in riverine flood risks. Compared to other countries, France, the Netherlands and the UK are projected to face the largest increases in exposure to risk of coastal floods, a driver to further increase investment in flood protection.

Based on analyses in Part IV, the degree to which EU member states are able to finance projected investment vary substantially. Some will face difficulties to raise tariffs for water supply and sanitation services without facing affordability issues for a significant part of the population (Bulgaria, Italy, Poland, Romania). Others are unlikely to find the fiscal space to increase public finance for water-related investment, as public budgets are constrained by public debt or high fiscal pressures (Belgium, France, Greece, Italy, Portugal or Spain). While few countries have experience with commercial debt to finance water-related expenditures, some of the countries facing the steepest increase in expenditure needs are, at the same time, the least likely to mobilise domestic commercial finance (Estonia, Latvia, Lithuania, Romania, Slovakia).

This final part of the report explores policy recommendations that can help meet financing needs. The recommendations are clustered around three sets of mutually reinforcing categories (see the table below):

- Make the best use of existing assets and financial resources
- Minimise future financing needs, and
- Harness additional sources of finance, where appropriate.

Table 5.1. Policy Recommendations to meet water-related financing needs in Europe

Make the best use of existing assets and financial resources	Minimise future financing needs	Harness additional sources of finance
Enhance the operational efficiency of water and sanitation service providers	Manage water demand	Ensure tariffs for water services reflect the costs of service provision
Encourage connections, where central assets are available	Strengthen water resource allocation	Consider new sources of finance
Develop plans that drive decisions	Encourage policy coherence across water policies and other policy domains (including nature-based solutions)	Leverage public and cohesion funds to crowd-in domestic commercial finance
Support plans with realistic financing strategies	Exploit innovation in line with adaptive capacities	
Strengthen capacity to use available funds		
Build capacity for economic regulation		

The recommendations are illustrated by good international practices. While tailored to countries facing most severe financing challenges, the recommendations are likely to be relevant for all EU member states and beyond. Preliminary discussions highlighted the potential benefit of peer-to-peer learning, possibly supported by technical support.

The chapter was informed by discussions on financing challenges and policy options during country workshops convened by the OECD and the European Commission in 9 EU member states facing most severe financing challenges: Bulgaria, Croatia, Cyprus, Lithuania, Poland, Romania, Slovakia, Slovenia, Spain. The discussion also builds on on-going analyses of similar issues at the Roundtable on Financing Water, an initiative by the OECD, the Netherlands, the World Water Council and the World Bank to accelerate water-related finance.

5.1. Options to make the best use of existing assets and financial resources

Improving the operational efficiency and effectiveness of existing infrastructure and service providers can postpone investment needs and is a prerequisite to further investment in water security. This can be enhanced through better operation and maintenance of existing infrastructure, demand management, and engagement with stakeholders (to set acceptable levels of service, enhance willingness to pay, or drive water-wise behaviour). Such a line of action resonates with discussions at country workshops convened in the context of this project in countries facing the most severe financing challenges to comply with the three technical directives discussed in this report.

The ensuing sections present policy insights and guidance for the following recommendations to make the best use of existing assets and financial resources:

- Enhance the operational efficiency of water and sanitation service providers
- Encourage connections, where central assets are available
- Develop plans that drive decisions
- Support plans with realistic financing strategies
- Strengthen capacity to use available funds
- Build capacity for economic regulation.

5.1.1. Enhance the operational efficiency of water and sanitation service providers

Operational efficiency of water service providers is a condition to make the best use of existing assets and financial resources. It is also a requisite to attract other sources of finance, be they public or private. It is essential to maintain or increase water users' willingness to pay for tariffs that reflect the cost of service provision.

Enhancing operational efficiency of service providers can take different forms depending on the national context. Building on international good practices (see OECD, 2018, for a discussion), performance indicators for water supply and sanitation services can focus on the following items. The relevance and relative weight of indicators would reflect local conditions:

- Technical performance indicators
  - Leakage performance and targets for reducing leakage and other unbilled losses, such as illegal connections
  - Mains bursts (as a proxy for distribution network condition)
  - Sewer collapses (as a proxy for sewer asset condition)
  - Number of wastewater pollution incidents, such as from too frequent operation of combined sewer overflows, or major failures at wastewater treatment works
  - Unplanned outages (loss of supply because of bursts, contamination, etc.)

- Compliance with existing regulation
    - Drinking water quality compliance (integrating with and reinforcing the role of the drinking water regulator, where this is separate)
    - Level of compliance with environmental permits and standards (integrating with and reinforcing the role of the environmental regulator, where this is separate). This can also be an indicator of the quality and state of drinking water and wastewater treatment infrastructure assets
- Customers' experience
    - Reducing per capita consumption for households and demand in other sectors on mains supplies
    - Risk of demand restrictions in a drought
    - Customer experience: how well billing queries are dealt with, information about planned outages and supply interruptions.

Analyses in this report and discussions at the country workshop indicate that Bulgaria would benefit from a proactive approach to maintain and renew existing networks (instead of reacting to incidents such as bursts) to improve operational efficiency of water and sanitation operators, reduce non-revenue water and address the backlog of under-investment in maintenance of WSS infrastructure over the past decades. This includes improving the operation of the existing assets to reduce operational costs and avoid additional capital investments. It also includes active leakage control in the water supply system and regular maintenance of pipes of the collection systems. Performance based contracts may be considered to strengthen incentives for investing in efficiency improvements. Technical assistance for service operators could include capacity building for financial and technical dimensions of operations[3].

In Poland, there has been rapid investment in infrastructure over the past years, but at the same time, a considerable part of the network is aging, which needs renewal and modernisation. Targeted maintenance, on a risk-based approach can help optimise spending, if data on the state of infrastructure is available.

In water-scarce countries like Cyprus, reduction of non-revenue water can minimise pressure on the resource and avoid (or postpone) investments in costly alternative water sources such as desalination. In Cyprus, this requires reducing leakage and increasing collection of water bills, especially under the jurisdiction of municipal water departments and community boards.

In Romania, reducing non-revenue water due to illegal connections (often associated with irrigation water use) and under-metering should be prioritized in the short term. In addition to generating more revenue for operators, addressing commercial losses should improve the official figure for the national piped water access rate, as currently un-registered connections would be taken into account. This process would be less time and cost intensive than addressing leakages in the distribution network.

Operational efficiency can benefit from benchmarking and public reporting of operators of water supply and sanitation services to increase accountability, transparency and incentives for efficiency and financial sustainability. In Lithuania, benchmarking has helped identify issues related to fragmentation and lack of efficiency of service provision for water supply and sanitation and could pave the way to reforming the sector. In Romania, benchmarking the performance of regional service providers on such indicators as leakage, reduction in illegal connections to networks, or number of staff per volume of water sold or treated could drive support for the needed efficiency gains.

Several countries (e.g. Cyprus, Lithuania, Slovakia) would benefit from exploring mechanisms to enable further consolidation of municipal and local services to improve operational efficiency and financial sustainability by reaching economies of scale. On-going reform in Croatia still needs to come to fruition, as progress has been slow. Experience from Hungary (where a staged approach has allowed a consolidation of utilities) or Ireland (where Irish Water was set up as a national service provider) can inspire other countries. Experience in Croatia and Romania shows that such institutional reforms can capture the attention of authorities and stakeholders, and delay other needed reforms. Planning, stakeholder engagement and sequencing reforms are essential to avoid capacity bottlenecks and overcome resistance to consolidation of utilities.

5.1.2. Encourage connections, where central assets are available

In countries such as Croatia, Lithuania, Romania and Slovakia, water users' reluctance to connect to existing central water supply and sanitation infrastructure delays progress towards compliance with the DWD and the UWWTD. In such contexts, connection to central supply and water treatment systems can be encouraged, possibly through regulation, with a direct subsidy to households to cover (parts of) connection fees or by allowing one-time connection fees to be paid in smaller increments over time.

Similarly, in Slovakia, connection to central sewer systems could be incentivised, to reduce costs of water pollution and drinking water treatment, and to provide a new source of revenue for water supply and sanitation utilities. Options may include:

- increased monitoring, enforcement and issuance of financial penalties for mismanagement of individual and other appropriate systems
- direct government subsidies building on the success of the "let's connect" programme (connection for EUR 1);
- incorporating the cost of connection into the overall capital cost; and
- public education and awareness on the environmental impacts of IAS and the consequences of inaction.

5.1.3. Develop plans that drive decisions

To make the best use of existing assets and financial resources, most countries would benefit from proper planning and priority setting. Investment planning should factor in demographic trends, including depopulation of rural areas and smaller towns to avoid over-investment in oversized infrastructure that will be costly to operate and maintain in the future. For example, the rural population is projected to contract by 40% in Romania in the coming decades, which has implications for current infrastructure development.

In non-viable areas, such as mountainous and isolated areas, cost-effective decentralised wastewater collection and treatment could be considered. Compliance monitoring and enforcement will be crucial to ensure environmental protection (i.e. to prevent groundwater contamination from leaking septic tanks, and inappropriate wastewater disposal without treatment to rivers).

In Cyprus, first order priorities include investments in sewerage networks, wastewater treatment and nature-based solutions that maximise benefits (to society and the environment) over the long-term and deliver the highest benefits in terms of compliance with the EU Urban Wastewater Treatment Directive and the Water Framework Directive. As compliance with the UWWTD increases, wastewater reuse could be expanded, where appropriate, to reduce pressure on groundwater resources.

Setting priorities can also contribute to cost-effective flood protection. In Slovakia, the 588 flood hazard areas should be reviewed to narrow the number of areas at highest risk and prioritise investment.

Effective plans must be consistent with initiatives in other sectors. For instance, in Cyprus, water management plans should be accompanied by a viable strategy for irrigated agriculture in line with sustainable aquifer management and the requirements of the EU Water Framework Directive.

Going beyond the compilation of individual projects, plans should consider how investments can be sequenced over time to improve resilience. This requires a shift from cost benefit analysis at project level to an assessment of the value created by investment pathways that combine and sequence a series of investments.

5.1.4. Support plans with realistic financing strategies

Plans and priorities should be accompanied with robust and realistic financing strategies. Such strategies are often lacking (e.g. in Cyprus, Romania or Slovenia) or pending (e.g. in Bulgaria). Strategies should clearly set priorities and drive investment decisions, and be developed in cooperation with national and local authorities. They should include provisions for improved operation and maintenance of water infrastructure, accounting for the backlog of under-investment in maintenance over the past decades. Strategies should also include targeted social measures to address affordability constraints and solidarity mechanisms to help cover investment costs in communities where financing capacities are especially limited.

The European Commission recently proposed an enabling condition to access further finding that goes into that direction. It remains to be seen how specific and comprehensive that condition will be.

The lack of a realistic financing strategy is especially acute for small municipalities (and rural areas). In Poland, for example, there is a mismatch between high investment needs, and technical and financial capacity of small municipalities (mainly rural). Affordability issues arise in smaller towns. Slovenia faces as similar situation.

In Bulgaria, due to the limited financing available and severe financing challenges in the future, there is a need to develop a consolidated vision of financing needs for compliance with the EU water *acquis*. This could ensure stronger policy coherence and alignment of priorities, as well as optimisation of limited resources use. The prioritisation of investments should systematically explore opportunities to combine funding to serve multiple objectives (water supply, flood risk management, pollution abatement, improving ecological status, etc.) to improve cost-effectiveness. Prioritisation should be considered in terms of policy objectives as well as geographies. Investment planning should factor in demographic trends, including depopulation of rural areas and smaller towns and economic trends (e.g. declining industrial use) to avoid over-investment in oversized infrastructure that will be costly to operate and maintain in the future. Priorities should reflect cost-benefit analyses made for RBMPs' programmes of measures, or explain why they do not.

In Cyprus, a sustainable financing strategy for operation and maintenance of water infrastructure is needed, accounting for the backlog of under-investment in maintenance over the past decades, in cooperation with national and local authorities. This should include ensuring that revenues collected from water and sanitation tariffs are sufficient to cover, and are earmarked for, operation of utilities (which does not seem to be the case in rural areas). Without such a strategy, delays in implementation of the EU water *acquis* and dependence on EU funding may continue.

5.1.5. Strengthen capacity to use funds effectively and financial disbursement at national level

Capacity to use funds effectively and financial disbursement play a critical role in allocating funding when and where it creates most value. Several countries face difficulties to invest available funds in an effective and efficient way. For instance, disbursement of cohesion policy funds has been delayed in Croatia and Romania. Delays can affect the robustness of project selection and implementation, or generate tensions with the Treasury who may be tempted to redirect available finance to sectors where it can be used effectively.

In such contexts, the capacity to use funds effectively should be strengthened. Along with general capacity building, this could be done through developing a strong project pipeline, and measures to ensure the sustainability of investments. Other issues will need to be addressed, which go beyond the water sector and the ambition of this report; for instance, cumbersome public procurement procedures in Slovakia, or labour shortage in civil works and construction industry, have delayed realisation of investments in Croatia.

In Slovakia, the efficiency of expenditure programmes could be enhanced. The Environmental Fund has a pivotal role to play. Currently, it only supports small projects, below EUR 200k. Revenue from new economic instruments (e.g. environmental fines or pollution charges) could be earmarked for the Fund to better support larger projects.

5.1.6. Build capacities for economic regulation

Independent economic regulation can support the transition towards sustainable financing strategies. Key features of well-defined independent regulation are to separate functions and powers of policy from operations, and to incentivise greater performance and accountability from local authorities, operators of water services and water users. Such oversight could provide technical support to local authorities, strengthen the transition to full cost recovery tariffs, and ensure consistency of tariffs across regions and communities (OECD, 2018; 2015d). Experience in the UK, and more recently in Lithuania, can inspire other countries where independent regulation is missing (e.g. Cyprus).

Poland has made important strides in the evolution of the legal and institutional framework for the sector, including tariff regulation. In December 2017 amendments in legislation governing water use (Acts on Collective Water Supply and Collective Sewage Disposal) established a new regulatory office to oversee water tariffs. The main aim of this amendment is to ensure tariffs are affordable, while also taking into account the financial stability of service providers.

Where national regulators do exist, they may need to be strengthened. This is the case for ANRSC in Romania. Regional utilities will progressively need to finance larger portions of their investment through revenue collection. ANRSC will need to enhance monitoring of operational efficiency, strengthen revenue-raising capacities and introduce proper incentives. An important consideration will be how to include depreciation of existing assets in the calculation of allowable tariff levels. This is an issue for a number of EU member states and there may be scope of joint action and peer learning on that front.

Further, barriers to the effective execution of economic regulation should be removed. For example, in Bulgaria, although the recent sector reforms sought to provide greater clarity on the allocation of roles and responsibilities between owners (Water Associations) and operators (Water and Sanitation Operators), overlaps and inconsistencies among territories remain. This has stalled the approval of operators' business plans by the regulator – the Energy and Water Regulatory Commission of Bulgaria (EWRC).

International good practices for water regulation

There are three core elements of water regulation:

- Protecting the environment: ensuring that standards are set and met in order to achieve policy objectives, and that abstractions and discharges operate within safe limits.
- Protecting customers' interests (economic regulation): ensuring that the delivery of water supply and sanitation is efficient, the level of charges fairly reflect and fund the quality of service delivered, and that there are equitable, transparent grievance and remedy mechanisms that allow individuals to complain.
- Protecting drinking water quality: providing confidence to customers that water treatment processes are effectively managed and monitored, and that tap water is safe to drink.

Independent regulation can be achieved by any one, or a combination of, the following four models (OECD, 2015a):

1. Regulation by government. The public sector is responsible for the management of the water services and owns the assets. Service provision is delegated to public water operators while regulatory functions are carried out directly by the State at different levels: central, regional or municipal. This is the model adopted in the Netherlands, and to a lesser extent, in Germany. The challenge for this regulatory model is that one public body is regulating another.
2. Regulation by contract. The regulatory regimes are specified in legal instruments, and although public authorities are responsible for regulation, water service delivery can be delegated to private operators through contract agreements. These set the rights and obligations for each contracting entity, and service provision is awarded to private companies following public tender. This model is used in France.
3. Regulation by one or multiple independent regulators, where independence has three dimensions: independence of decision making, of management and of financing. This is the model used in the United Kingdom, where the regulatory framework is organised around three dedicated agencies with statutory functions relating to pricing and customer service (Ofwat), drinking water quality (Department for Environment, Food and Rural Affairs), and environmental regulation and security of water supply planning (UK Environment Agency).
4. Outsourcing regulatory functions to third parties. This model makes use of external contractors to perform activities such as tariff reviews or benchmarking.

Regulators sit between government and its policy making, the bodies responsible for the delivery of water supply and wastewater services, and their customers. This means that they must translate government policy aims into operational standards for those whom they regulate.

How a regulator acquires performance information and sets performance targets is important in bridging any gap between government and customer expectations. An outcome-based approach helps to ensure that the focus is not simply on easily measured outputs, but also considers the longer-term aims for water and sanitation, and the environment. It should expect the delivery body to monitor its service to customers, the operational performance of its assets, and how it is planning for resilient systems operation in the face of shocks, such as drought, process failures or cyber-attacks. The targets, and performance against them, should be published and made available to customers.

Customers should expect to be able to express their views on levels of service, priorities for investment and options for major infrastructure where this is proposed. The extent to which customers participate in the development of business plans can influence both their behaviour – and how much they value water and the service they receive – and that of the delivery body.

The regulator needs to ensure that the delivery body is funded to deliver efficiently the breadth of its services to the required standard. For household water and sanitation bills, affordability issues are best dealt with through the use of social tariffs or income support measures (outside of the water bill), rather than keeping water bills low and failing to raise adequate revenue and an understanding of the value of water and sanitation services. The United Nations has stated (UNESCO, 2017) that regulatory frameworks must not interfere directly or indirectly with people's existing access to water and sanitation. States must ensure that disconnections due to inability to pay are prohibited. The regulator should seek to moderate bill increases so that it can satisfy itself, and others, that they are necessary and appropriate.

5.2. Options to minimise future financing needs

Options discussed in the previous section contribute to making the best use of existing assets and financial resources. They also minimise investment needs in the future, for instance by postponing the need to renew existing infrastructures. The ensuing sections present policy insights and guidance on the following additional measures to minimise future financing needs:

- Manage water demand, and strengthen water resource allocation
- Encourage policy coherence across water policies and other policy domains
- Exploit innovation in line with adaptive capacities.

5.2.1. Manage water demand

Water demand management can go a long way to minimising future needs to invest in supply augmentation. In Cyprus, demand management efforts (illustrated by the recent tariff reform, abstraction charges, awareness raising campaigns) can be scaled-up to reduce the need for costly supply augmentation. A revision of the WSS tariff structure and level can contribute to drive water use efficiency, with a higher proportion of – or rate for - volumetric charges (and a lower proportion of fixed charges), especially in small communities.

Abstraction charges in most countries are typically low or non-existent. However, freshwater abstraction charges to all users can signal the value of water and limit the pressure on water resources, particularly groundwater. Groundwater and surface water abstraction charges should be set in a manner coherent with each other, to account for potential substitution effects. When water abstraction is metered, a volumetric charge should be applied. If abstraction is unmetered, a flat abstraction charge or one based on a proxy, such as area of irrigated land (preferably in conjunction with the type of crop), can be used as a more rudimentary alternative in the interim (Ambec et al., 2016). The price should reflect the trade-off between abstracting water now or in the future, particularly for non-renewable groundwater resources (OECD, 2017). The revenue raised could be earmarked to fund water restoration activities. In addition, collection of water bills, particularly for unregistered abstractions and dealing with illegal abstractions should be a priority in some countries, such as Cyprus, to manage demand and ensure sustainable water abstractions.

5.2.2. Strengthen water resources allocation regimes

Several countries which face severe financing challenges could benefit from well designed water allocation regimes. For instance, in Cyprus, much water is allocated to low value agriculture uses, driving costly investments in supply augmentation and depleting the resource, thus weakening compliance with good ecological status. A reform of water allocation regimes would contribute to water use efficiency, discourage wastage and low value uses and secure water for valuable ecosystems. It should account for the fact that, while the share of agriculture in GDP declines, agriculture still is important for the national economy in terms of social cohesion, countryside and local tradition, and employment.

In Spain, water allocation regimes have led to costly reforms to buy back entitlements to contribute to environmental flows and to allocate water towards higher value uses. Current practices also led to costly investments in supply augmentation via desalination, a direction criticised by the Court of Auditors.

OECD (2015) suggests that an allocation regime needs to have two key characteristics: it should be robust by performing well under both average and extreme conditions and demonstrate adaptive efficiency in order to adjust to changing conditions at least cost over time. More specifically, a well-designed allocation regime has multiple elements (discussed in detail in OECD, 2015b). A clear legal status should be in place for all types of water resources (surface and ground water, as well as alternative sources of supply) with competing claims clarified. A clear and enforceable abstraction limit ("cap") should be in place that accounts for in situ requirements and sustainable use, including environmental needs. Clearly defined, legal, volumetric water entitlements are needed. Water pricing, typically in the form of abstraction charges, is a key element of a well-designed regime. Pricing can contribute to cost recovery, internalise negative externalities associated with water abstractions, and send a price signal to users to discourage inefficient and low-value water uses. Scarcity pricing could help to signal the scarcity value of the resource, but has proven difficult to implement to date.

Groundwater allocation faces distinctive challenges, for instance as regards abstraction monitoring. However, a similar reasoning applies (see OECD, 2017 for a more detailed discussion).

Recognising the potential for improving current allocation arrangements, 75% of countries covered in the OECD survey have recently reformed their allocation regimes and 62% have reforms ongoing. However, managing the transition from existing arrangements to an improved regime is often very contentious and can be costly. Evidence from case studies of allocation reform in 9 OECD countries and BRIICS (Australia, Canada, Chile, China, France, Israel, South Africa, the UK, the US) provides insight into the reform process and lessons on how some of the obstacles of reform can be overcome.

Concerns about water scarcity and insufficient water for ecosystems are often cited drivers of allocation reform. Broader political or structural reforms have provided imperatives to improve the efficiency of resource use and equity in allocation of water resources. Droughts can provide a salient, visible event to trigger action. The case studies on reform highlight the importance of determining a sustainable baseline (how much water is available for allocation) before making significant changes, like introducing trading. Failure to do so can result in costly efforts to claw back entitlements already granted. Willingness to engage stakeholders in the reform process and appropriately compensate potential "losers" (with financial transfers, permits to build storage structures) facilitates the process.

A periodic "health check" of current allocation arrangements can help to assess the achievement of reforms and areas for further improvement. The OECD "Health Check" for Water Resources Allocation can provide useful guidance for such a review (see OECD 2015b, 2017). It is a tool designed to review current allocation arrangements to check whether the elements of a well-designed allocation regime are in place and to identify areas for potential improvement. In general, as the risk of shortage increases, the benefits of a more elaborate allocation regime increases.

5.2.3. Encourage policy coherence, across water policies and other policy domains

Policy coherence can contribute to minimising future financing needs. Coherence between water supply, sanitation and flood protection with agriculture policy is a case in point, as it could deliver a number of significant co-benefits. While the regulatory framework is largely set at European level, implementation varies across countries. In addition, countries would benefit from exploring flood management options that exploit synergies with the Water Framework Directive, such as flood protection measures that minimise alteration of the natural hydromorphology. From that perspective, nature-based solutions can reinforce and support existing flood defence investments (see further developments below on nature-based solutions for flood protection) while preserving the good ecological status of water bodies.

In Cyprus, improved coordination between land use planning and flood management could manage urban sprawl and flood hazards, and minimise investment needs for flood protection. In Slovakia, it is advisable to review and adjust the ten sub-basin water management plans so as to increase synergies between policies (including those for agriculture, water supply and sanitation, water quality, flood prevention, land use planning, nature conservation and climate change adaptation) and the objectives of the EU water acquis. Also, flood management could be decentralised to the local level to better reflect local priorities. In addition, greater emphasis could be placed on integrating flood prevention into river basin management plans, on better use of nature-based solutions, and improved coordination between land use planning and management and flood prevention.

Nature-based solutions in urban environments

Nature-based solutions (NBS) involve the use of natural or semi-natural systems that utilise nature's ecosystem services in the management of water resources and risks. Examples include restoration or construction of wetlands, sustainable urban drainage design and green roofs. NBS can contribute to addressing water-related challenges in urban environments, often supporting multiple policy objectives (see Table below). In the European context, NBS are particularly relevant for catchment protection and protection against flood risks.

Table 5.2. Benefits of nature-based solutions versus traditional engineered 'grey' infrastructure for urban water management

Green infrastructure solution	Urban water management issue						
	WSS (including drought)	Water quality regulation			Moderation of the extreme events (floods)		Protection of ecosystems
		Water purification	Biological control	Water temperature control	Reverine flood control	Urban stormwater runoff	Coastal flood (storm) control	
Demand management	x							x
Local processing of black or grey water	x	x	x					
Wetlands restoration/conversation	x	x	x	x	x			x
Constructing wetlands	x	x	x	x	x			x
Water harvesting	x					x		
Green spaces	x	x		x		x		x
Permeable pavements	x	x				x		x
Green roofs						x		x
Protecting/restoring mangroves, coastal							x	x
Corresponding grey infrastructure (primary service level)								
Dams, groundwater pumping	x				x			
Dams, levees					x	x		
Water distribution systems	x							
Water treatment plant		x	x					
Urban stormwater infrastructure						x		
Sea walls							x	

Source: adapted from UNEP (2014), OECD (2013a).

Traditional approaches to urban drainage have historically focussed on fast, proven and safe water removal using engineered grey infrastructure. They provide various benefits, including flood control and the treatment of contaminated urban runoff. However, nature-based solutions for stormwater management provide many other additional benefits, beyond what equivalent traditional grey infrastructure provides. These benefits include, inter alia, adding to the stock of green infrastructure and natural capital and hence providing opportunities for improved ecosystem services and biodiversity; reducing ambient air pollution (Pugh et al, 2012); and mitigation of urban heat island effects.

Investment in NBS is also generally less capital intensive; has lower operation, maintenance and replacement costs; avoids lock-in associated with capital intensive grey infrastructure; and appreciates in value over time with the regeneration of nature and its associated ecosystem services (as opposed to the high depreciation associated with grey infrastructure). In the context of a changing climate, where rainfall patterns, water availability and demand are becoming more uncertain, nature-based solutions can provide a flexible, scalable option to adapt. Nature-based solutions can also avoid or postpone the costs of building new, or extending existing, grey infrastructure. This resonates with the guidance of the European Commission on the topic, which claims that a network of healthy ecosystems often provides cost-effective alternatives to traditional infrastructure, offering benefits for EU citizens and biodiversity. This is why the EU Strategy on Green Infrastructure promotes the use of nature-based green and blue infrastructure solutions.

Many countries have shifted their perspective to better reflect the net value of a policy or project to society as a whole, and also to promote the use of natural systems, such as nature-based solutions (e.g. Xing et al., 2017). Cities are gaining experience with these approaches (see the case of Philadelphia in Box 5.1).

Box 5.1. Nature-based solutions to manage stormwater in Philadelphia, USA

Philadelphia, USA is considering a range of nature-based solutions to adapt to a changing climate and mitigate urban floods from heavy rains. Ongoing stormwater enhancements in Philadelphia are bringing some USD 2.6 billion (approximately EUR 2.35 billion) of added benefits by managing combined sewer overflows using nature-based solutions that deliver multiple functions.

The proposed eco-friendly "sponge-like" water system in Philadelphia involves new forms of drainage: green roofs, wetlands and repaving with porous materials. It is estimated to be less than half the cost of a conventional upgrade of the current system of pipes and basins. Achieving a similar level of service through an additional wastewater treatment plant would be 4- or 5-fold more expensive.

Sources: Various; adapted from OECD (2015c), Water and Cities, OECD Publishing, Paris.

Although the benefits of green infrastructure for water security and sustainable growth are known, significant barriers hinder investments in green infrastructure:

- Due to limited data on river flows, as well as of evidence on the value of freshwater and terrestrial ecosystems, it can be difficult to design and assess the costs and benefits of NBS. The estimation of benefits is especially complex as these may be hard to quantify and monetise. Much of the current efforts are directed towards developing the means of financial valuation of aspects of improvement in urban areas that are not readily captured in terms of market value[4]. As a consequence, there is a lack of track-record for costs and benefits, technologies, markets and financial products associated with NBS. The absence of available best practices and expertise for investors creates uncertainty related to bidding processes, timing for investments, transactions and underlying risks.

- NBS use vast areas of land, which can be expensive especially in urban contexts. However, once benefits are taken into account, the overall net present value of using NBS can be better than the traditional approaches. The major challenge is that many of the benefits do not accrue to water services or water service providers, instead going to elevated property values, health benefits and other non-water related societal benefits (e.g. Zhou et al., 2013; Ashley et al., 2018).

- NBS are multipurpose by nature, thus contract structures may be highly complex and vulnerable, exposing investors to risks and insecure returns.
- The innovative practices associated with NBS often combine different scales in urban water management, from individual buildings, to municipal and larger levels. Such combinations can be hampered by institutional arrangements, which split incentives and responsibilities along the water cycle.
- There are certain liability issues that cannot be addressed in the case of NBS, due to the intrinsic characteristics of its components, which rely on natural ecosystems. There is great ambiguity related to the determination of who to hold accountable in case of failure, for example, when a floodplain ceases to deliver the services it is expected to provide.
- Many investors have yet to conclude that green infrastructure investments offer a sufficiently attractive risk-return profile. A number of environmental, energy and climate policies and regulations still favour investments in grey infrastructure over green infrastructure. The competitiveness of innovative solutions is often hampered by lack of policy coherence; for instance, water prices that fail to reflect the opportunity costs of resource use; or land use and urban development that do not reflect the risks of building in flood plains. There is a need for support policies and funding mechanisms that price nature-based and ecosystem services in ways that encourage investments in green infrastructure.

The abundant literature and documentation of successful case studies (e.g. OECD, 2015) can inspire policies in EU member states. Issues remain regarding replicating or scaling-up experience with financing for NBS that contributes to water security and sustainable growth.

Nature-based solutions for flood protection

In the past, efforts have been made in Europe to control flooding. Traditional flood defence measures, also referred to as grey infrastructure measures, have been implemented on a large scale. They are defined as: "Grey infrastructure refers to man-made infrastructure. In the context of floods, it refers to dams, dikes, channels, storm surge defences and barriers in general. It is called 'grey' because it is usually made of concrete" (EEA, 2017).

Even though traditional flood defence measures have proved to be effective, it has become increasingly recognised that they actually work against nature and negatively affect the provisioning of ecosystem services. The financial and ecological challenges have pushed forward the search for more sustainable flood protection solutions that work with nature and contribute to the strengthening of the resilience of nature and society to flooding. The European Commission defines green infrastructure as a particular nature-based solution that can be applied to flood risk management and is defined as: "A strategically planned network of high quality natural and semi-natural areas with other environmental features, which is designed and managed to deliver a wide range of ecosystem services and protect biodiversity in both rural and urban settings." (EC, 2013).

In the context of flooding, green infrastructure contributes to minimising flood risk, by using ecosystem-based approaches for flood protection, for example through flood plain restoration rather than constructing dikes only. Thereby, green infrastructure does not only provide benefits to society in the form of avoided flood damage like grey infrastructure. Nature-based flood protection measures provide additional benefits, such as biodiversity improvements, water quality improvements, or opportunities for recreation.

NBS to protect against river flooding can be classified into agricultural, urban, hydromorphological and forest categories (NWRM, 2016). Agricultural measures are mainly aiming to manage run-off and thereby reduce flood risk, for example through soil and land management measures (conservation tillage, early sowing, hedges, buffer strips), drainage measures (flow-paths in fields, controlled traffic farming, reduced stocking densities) and run-off pathway measures (farm ponds, swales, mulching). Forest measures aim to manage woodlands in such a way that they reduce flood risk by intercepting land overflow, or by encouraging infiltration and soil water storage. Hydromorphological infrastructures include wetland restoration and management, floodplain restoration and management, re-meandering, and stream bed re-naturalisation.

NBS that can be considered to protect against coastal flood risk include saltmarsh and mudflat management restoration, sand dune management, and beach nourishment. Advancing nature-based solutions for flood management often boils down to a discussion on the costs of NBS and whether or not these measures provide sufficient benefits in comparison with traditional grey alternatives.

The discussion of the comparative advantages and disadvantages of NBS and grey infrastructure measures focuses on the following criteria (EEA, 2017):[5]

- Costs of land acquisition and compensation: The financial expenditures incurred from buying land needed for the construction of the measure or compensation of landowners for externalities associated with the construction of the measures.
- Construction and rehabilitation costs: All costs incurred during the construction phase of the measure including the investments in equipment, infrastructures and other assets required as well as the associated labour and management costs.
- Operation and maintenance costs: Financial costs such as depreciation allowances, maintenance expenditures and operational expenditures.
- Effectiveness on flood protection through:
  - Capturing the features of water retention relating to storing surface run-off and slowing run-off through slowing movement of surface water without storage.
  - Capturing the features of water retention relating to storing or slowing river water through e.g. open or controlled connectivity of plains or increasing bed roughness.
  - Reducing runoff through increasing evapotranspiration, increase infiltration and/or groundwater recharge and increasing soil water retention.
- Side effects: All benefits or costs related to flood protection measures in addition to the initial flood protection objective itself.

In the evaluation of NBS and grey infrastructure measures, comparing costs and effectiveness (in a cost-effectiveness analysis) is not a fully informative exercise. The key advantage of nature-based solutions is that they deliver multiple benefits: in addition to flood protection they often provide a wide range of other ecosystem services (i.e. on water quality and recreation). In other words, NBS might be less cost-effective, but more cost-efficient, once all different types of benefits are taken into consideration. In general, the cost-efficiency hinges on land acquisition costs and additional benefits from ecosystem services.

It is difficult to generalise and to draw conclusions about the return on investments of NBS for flood risk management in comparison to traditional grey alternatives as the magnitude of costs and benefits highly depend on location specific characteristics. Either set of measures can have a relative advantage depending on the situation (geographical context, population density, economic activity, local price levels, and size of the project) to which it is applied. So, it is misleading to compare green versus grey infrastructure measures on a one-to-one basis (EEA, 2017). Table 5.7 summarises the overall advantages and disadvantages of NBS and grey infrastructure for flood protection, based on several criteria.

Table 5.3. Overall comparison of NBS and grey infrastructure for flood protection

	Grey infrastructure	Nature-based solutions
Effect on flood risk	Protective: they protect an area against damages from flooding by blocking water from passing into a specific area.	Preventive: they reduce the probability of a flood occurring.
Potential to withstand extreme flood events	Yes, measures are designed according to a certain protection standard with established monitoring techniques.	Uncertain, effectiveness to withstand extreme flood events not demonstrated. NBS is often implemented together with grey infrastructure to guarantee protection for extreme events (also called hybrid infrastructure).
Timing of functionality	Immediate, grey infrastructures provide service as soon as operational.	Delayed, NBS typically needs time to provide the service as rivers/coasts adjust their morphology in response to the measures. This process depends on physical processes that take time, causing a delay in reaching flood protection standards.
Land acquisition and compensation costs	Land acquisition and compensation costs are low due to a small physical footprint.	High physical footprint, consequently land acquisition and compensation costs are high (except for re-naturalization). Costs depend on actual land-use (agriculture, residential) and the market value of land in the area.
Construction and rehabilitation costs	Need for recapitalization: depreciating assets that need regular upgrading or replacement.	Recapitalization is not significant, these measures are often self-sustaining and do not depreciate.
Operation and maintenance	High	Low operation and maintenance costs. NBS are often to a large extent self-sustaining and reduces the need for maintenance compared to channelized rivers.
Ecological footprint	Increased ecological footprint due to the use of unsustainable materials and energy intensive processes.	Low ecological footprint due to the use of natural materials and processes.
Side-effects	Limited positive side-effects on the provisioning of ecosystem services. The protective nature of grey infrastructure has a potential perverse incentive on location behaviour by attracting residents and economic activity to flood-prone areas (thereby increasing flood risk).	Provisioning of a large range of ecosystem services

Source: EA, 2017.

5.2.4. Exploit innovation in line with adaptive capacities

Innovative technologies and management systems are being developed, which provide opportunities to reduce investment needs in water supply and sanitation, and in flood protection, now and in the future, in particular in the context of a changing climate. EU Member states already equipped with infrastructures may face distinctive challenges to transition towards alternative systems: technical path-dependency and risks of stranded assets can limit the appetite for and the feasibility of alternative systems, at least in the short term. Member states where additional infrastructure is required may find it easier to adopt alternative systems and techniques and ultimately perform better with less capital costs.

This section focuses on technical innovation, with a zoom on distributed systems for water supply and sanitation. Innovation does not come in isolation; innovation delivers best when combined with financial and governance measures, and when the interface between urban and rural environments is properly addressed. For example, sustainable urban planning, water-sensitive urban design, innovative business models and dedicated policies to drive innovation can all minimise future financing needs.

Innovation and compliance with the EU acquis on water

Innovation can minimise the costs of water management. Water-related innovation is multifaceted:

- In agriculture, innovation is associated with the development of water-efficient irrigation, planting of less water-intensive crops, and the adoption of practices that reduce nutrient flows back to water bodies.
- In manufacturing, it deals with more water-efficient and cleaner production practices, appliances, and more effective treatment techniques. Similar opportunities are associated with water supply and sanitation.
- Innovation applies to storage techniques, monitoring of river flows and pollution loads, and the operation of infrastructure as well. Smart water technologies cut across these boundaries: they allow the users to monitor, manage and act on data relating to the part of the water cycle that is pertinent to their interests.
- Data management advances, associated with optimisation of monitoring systems, will lower costs of both demonstrating compliance and operating water service systems. Online and dynamic real time data will become more readily available for flows, pollutants and quality of water at taps.

Table 5.7 presents a summary of technologies and innovations for water systems, and social costs in comparison to traditional water systems.

Water-related innovation is not limited to new technologies: non-technical innovations can also contribute to water security and sustainable growth. Sustainable urban planning is a good illustration. The way in which towns and cities are planned and laid out in terms of building form, density and surface transport networks, has for a long time controlled the form and layout of water services. The buildings and roads are planned and designed, with the assumption that the water services will be added appropriately, usually using below-ground pipes laid beneath roads, conveying water into the area from upland treatment works and wastewater away to downstream and remote wastewater treatment plants. The surface water flows into highway and building drains, to join with the sanitary wastewater and thence to the treatment plant, unless the pipe capacity is exceeded and overflows occur.

Water-sensitive urban design takes an opposite perspective: it factors water-related risks into the design and planning of urban developments. As synthesised by van der Brugge and de Graaf (2010), water-sensitive urban design encompasses all aspects of urban water management, with additional urban design principles; it affects the volume and form of water-related investments: store or use water on site, rather than rapid conveyance of storm water; capture and use storm water as an alternative source of water, thus demanding less potable water supplied by a utility; use vegetation for filtering purposes; use landscapes to protect water-related environmental, recreational and cultural values; harvest water in decentralised systems for various uses; and treat wastewater in decentralised systems. Water sensitive urban design is most appropriate in expending cities, relatively low density urban environments, or in countries or cities projected to be affected by heavier rains, due to climate change.

Innovative business models for water utilities are other good examples of non-technical innovation. The revenues of most water utilities depend on the volume of water sold and of wastewater collected and treated. There are benefits in (at least partially) decoupling revenue from the volumes of water sold. This can be done through the development of well-designed water tariff structures, and opening up opportunities to derive additional revenue by enhancing environmental performance through performance-based contracts (where the utility receives a premium when it reaches certain level of performance regarding, for instance, leak detection, or the quality of effluents). Such options could be considered, in particular in countries and cities where water use per household is projected to decline.

Water-related innovation may derive from dedicated policies. For instance, several countries and states (Arizona, Australia, California, France, Israel, Korea, Malta, the Netherlands, and Ontario) have explicitly encouraged the development and deployment of smart water systems, either to address local issues, or to support a growing global business (OECD, 2015c). The challenge is to foster country collaboration and transfer innovations to less-developed economies.

Innovation in urban water management: the case of distributed systems for water supply

Biggs et al. (2009) define distributed systems as a model where infrastructure and critical services are positioned close to points of demand and resource availability, and linked within networks of exchange. Services traditionally provided by a single, linear system are instead delivered via a diverse set of smaller systems - tailored to location but able to transfer resources across wider areas. According to the authors, distributed systems represent a localised and highly networked approach to production and consumption, and blur the line between centralised and decentralised water models: the central infrastructure plays an arterial role at a regional level, while smaller, tailored systems operate and interact with users at the local level.

Distributed water supply systems have been advocated as part of the 'soft path' for water, characterised by Peter Gleick as an attempt to "improve the overall productivity of water use and deliver water services matched to the needs of end users, rather than seeking sources of new supply" (Gleick, 2002). In a systematic review of cases from Australia, Europe and the US, Biggs et al. (2009) show how distributed water systems can generate positive outcomes that enhance and supplement those provided by existing infrastructure models. Distributed systems can:

- Reduce costs and resource use, by adapting water management to context and making the most of available resources.
- Improve service security and reduce risk of failure, by building redundancies in the system.
- Adapt to shifting conditions and demands and respond to risk and uncertainty, by increasing the diversity and flexibility of water systems without locking utilities, customers and future governments into rigid pathways for delivering critical services.

Distributed systems can be well-adapted for the transition from oversized to more adequate infrastructure (see OECD, 2013b). In some German areas, demographic decline combines with the decrease of per capita consumption to induce such collapse in water demand that public systems end up being largely oversized. Some public operators admit that it will not be possible to sustain the present infrastructure; since it would need rebuilding anyway, one option is to redesign them with room for distributed technologies at single family, block or community level, in particular at the urban fringe.

Table 5.4. Technologies and innovations for water systems

Technology/ innovation	Applicability	Examples	Implications for societal costs compared with baseline traditional systems	
			Existing assets (short to medium term)	New assets (medium to longer term)
Data, automation and control – real time monitoring, information and control (ICT)	Across the entire range of water systems, services and associated systems interacted with. There is a significant opportunity to maximise the potential of existing assets by getting the full use from them, by e.g. utilising the entire storage available by monitoring the volume and controlling flows, volumes and quality in real time.	Zimmer et al (2018) consider the use of predictive model based controls to manage combined sewer overfows spills for the Chicago deep tunnel sewer system. The added costs for the control options are compared with the avoided penalty costs for pollution incidents. Data analytics are now emerging as a concept for intelligent water systems control, enhancing efficiency and saving costs (e.g. Beach et al, 2018).	Overall costs to society should fall, although costs to service providers will increase in the short to medium term.	As for existing assets, but longer term system provision costs will fall for both society and operators.
Novel treatment processes	Especially relevant for wastewater systems in order to recover resource from waste and for different quality water supplies.	van Leeuwen et al. (2018) illustrate considerable opportunities to recover resources from waste that are now being exploited using novel technologies.	As most new processes will need to be retrofitted to existing treatment plants there will be added costs in the short to medium term.	There will still be added costs compared with the baseline in future, however, the benefits of e.g. resource recovery will far outweigh these.
Decentralised water systems	The emerging model of decentralised water service provision, using contemporary technologies.	The relative costs of centralised and decentralised systems depend on a number of factors (e.g. OECD, 2015). Institutional and operational arrangements are crucial to the optimisation of costs.	Where existing system provision is being expanded, decentralised systems may provide costs savings, but this will depend on context.	Decentralised systems can be installed at lower costs, depending on context, however, economies of scale will no longer be available.
Dynamic/intelligent asset management	This overlaps with the ICT category above and applies especially to (primarily capital) asset management	The extensive amounts of data now being collected routinely will mean that existing systems can be operated more effectively and even to extend the functionality beyond the original purposes.	There are significant ongoing added costs of installing monitoring and control systems into existing assets.	It may be that the clever data acquisition systems will be less costly in the future and a by-process of other data management systems, not water.
Multi-functional storm water/water systems	Systems that do more than simply provide safe water and safe wastewater disposal. Surface water in particular will provide opportunities for urban form and interact with lifestyles and willingness to accept more local systems	Water Sensitive Urban Design is an integrative approach to managing surface water more effectively in urban areas. Specific technologies will be required to do this using blue and green infrastructure and new modelling tools (e.g. Kuller et al., 2017). Lifestyle implications will need to be managed to ensure acceptability and uptake.	Taking advantage of the multi-functional opportunities that managing water differently can provide is going to cost more in the short to medium term as it requires retrofitting.	In the medium to longer term, multifunctional systems will become the norm, providing a wide range of added societal benefits. Although some costs will increase, these are likely to accrue to other societal service providers and individual property owners.

Technology/ innovation	Applicability	Examples	Implications for societal costs compared with baseline traditional systems	
Mainstreaming opportunities	Mainly applies to piggy-backing on the provision of others systems or services	Rijke et al., (2016) illustrate how mainstreaming can save costs when responding to the threats from climate change. Although so far focused on flood risk management and potentially drought management, mainstreaming could provide many opportunities for co-provision of water and other services.	The provision of water related benefits will cost less when added on to other services.	As multifunctional systems become more widespread, needs to mainstream will be less required.
Behavioural changes, and lifestyle opportunities	Interactions with the ways in which people live in urban areas can provide novel uses and options for water management, especially uses of different grades of water. This requires influencing behaviours to e.g. use less water, use different types of water.	Demand management has been effective in many parts of Europe, although the lack of application of full cost recovery continues to hamper individual restraint in the domestic sector.	Demand management has had mixed success. Full cost recovery may make the business case for different water uses and sources more compelling to domestic consumers. Overall societal costs may fall, although costs to service providers may increase.	There may be more opportunities to influence behaviour longer term, although to ensure this will require compelling regulation.

Note: green: lower social costs; red: higher social costs; yellow: no change or not applicable
Source: Various. Compiled and synthesised by Richard Ashley for this report (Ashley et al. 2018).Options to harness additional sources of finance

Most EU countries would benefit from exploring options discussed in previous sections, to make the best use of existing assets and financial resources, and to minimise future financing needs. These options can contribute significantly to closing the financing gap, in particular in countries where this gap is widest. Still, additional finance will be required to close the gap. This section explores options that can mobilise new sources of finance available in all EU member states. Success in mobilising these sources depends on progress made in the other two sets of options discussed above, which can be seen as requisites to harness additional sources of finance.

Policy options discussed in this section are particularly relevant in countries that depend the most on EU funding (see Chapter 2.1 above). Such dependence is detrimental to sustainable financing strategies at country level, as these funds are programmed to decline and be gradually phased out. The ensuing sections present policy insights and guidance on the following options to harness additional sources of finance:

- Ensure tariffs for water services reflect the costs of service provision
- Consider new sources of finance
- Leverage public and cohesion funds to crowd-in domestic commercial finance.

5.2.5. Ensure tariffs for water services reflect the costs of service provision

In most EU countries – and in particular in those facing the most severe financing challenges - there is room to ensure that tariffs for water supply and sanitation services reflect the costs of service provision. Further increases of WSS tariffs can ensure adequate funding for WSS service providers and control water consumption. For instance, Bulgaria is committed to move towards full implementation and enforcement of recovery of costs of water services, including environmental and resource costs. Room of manoeuvre is significant in Cyprus, where urban water supply and sanitation tariffs (both from water boards and municipal departments) are lower than in most other European countries (Marin et al., 2018). In Croatia, affordability concerns create very little room to manoeuvre in terms of tariff increases. One option would be seasonal tariffs in touristic areas, matching peak demand.

A usual barrier to tariff increases, affordability concerns are best addressed outside of the water bill, through targeted social measures. This is supported by a vast and convergent literature. The forthcoming OECD report *Addressing the Social Consequences of Water Tariffs* (OECD, 2020) explains how sophisticated tariff structures – including increasing block tariffs, which are increasing widespread across Europe - require a lot of information, which may not be available, and end up being regressive: they benefit households who could afford to pay more for water services, while depriving service providers from needed revenues. In so doing they hinder the extension of water services to deprived areas, hurting the poor or vulnerable groups.

In countries where a significant share of the population faces (or is projected to face) affordability issues, accompanying social measures may need to be supplemented by well-designed solidarity mechanisms at a larger scale. These may take the form of cross-subsidies across water users or territories (from urban to rural areas). Aggregation of services providers or organising authorities can facilitate such transfers. This is the case in Romania and Bulgaria, and to a lesser extent in Lithuania, Latvia and Poland. Croatia is considering moving into that direction.

5.2.6. Consider new sources of finance

Most countries would benefit from considering new economic instruments to raise additional revenue for water management and internalise pressures on water bodies (that result from abstraction or pollution). This may include the introduction of fertiliser and pesticide taxes to reflect the costs of water pollution, storm water taxes on property developers for impermeable surfaces that increase the risk of urban flooding, and payment for ecosystem services from utilities to farmers in exchange for the protection of catchments and the quality drinking water sources. Storm water taxes on property developers can raise revenue for flood protection measures and incentivise nature-based solutions, such as sustainable urban drainage systems. They may be relevant in Cyprus, for example, where the construction sector is dynamic.

Countries could also exploit synergies and combined investment opportunities with other sectors (e.g. urban development, food security, energy security, tourism) that reduce water-related risks. Options to align incentives through insurance schemes and land value capture mechanisms (such as local taxes on property value) should be explored on a case-by-case basis.

> **Box 5.2. Land value capture – a suite of tools to finance water-related investments**
>
> According to the "beneficiary pays" principle, expressed in the Vancouver Declaration during Habitat I (UN, 1976), the beneficiaries of public investments that valorise their land should partly cover such costs or return their benefit to the public. The means by which beneficiaries can pay back include taxes, such as land taxes and betterment charges; development charges or permit fees; pricing and compensation policies; adequate assessment of land values; and leasing publicly owned land (UN, 1976).
>
> Land value capture techniques can foster local urban development. Because public investments and planning decisions on urban development concern land in a very specific, localised manner, land value capture tools are a matter for local governments. Local governments may influence the direction of these projects to ensure the alignment with urban development and spatial planning goals.
>
> Land value capture tools can fund a wide range of urban development projects. Even though they are not associated with any particular type of investment, some projects could particularly benefit from the adoption of such tools, such as urban renewal projects and Transit Oriented Development (TOD) projects, given the great potential to trigger land valorisation.
>
> Experience in water-related projects is limited so far. Casablanca, Morocco, paved the way. Casablanca is characterised by rapid urbanisation; its population is expected to grow from 3.5 million to 5 million inhabitants by 2030. Extending the water network, securing access to the resource and protecting it against frequent floods are serious concerns for the local authority, which needs to finance these projects.
>
> The city defined a new investment programme in 2007. Revenues from user tariffs cover operational and maintenance costs and the renewal of existing assets (accounting for 70% of total cost over the last decade). A dedicated account (*fonds de travaux*) covers the remaining costs (essentially land acquisition, network extension and social connections). Financed mainly by contributions from property developers, it has financed a growing share of total investment, from 7% in 2004 to 54% in 2014. Property developers also cover the costs of connecting to the network and in-house equipment. Their contribution varies depending on the type of housing (social housing, villas, hotels and industrial zones), and they pay additional costs for developments that do not feature in the master plan. Contributions are waived when the developments take place in underprivileged neighbourhoods and slums.
>
> Source: OECD (2019), Land Value Capture: Framework and Instruments, unpublished paper; OECD (2015c), Water and Cities, OECD Publishing, Paris.

5.2.7. Leverage public funds (incl. EU) to crowd in domestic commercial finance

Domestic commercial finance is available across EU member states. As documented above, few countries have gained experience mobilising it for water-related expenditures. There is room of manoeuvre to attract commercial capital for creditworthy borrowers to finance water-related investments. This may require exploring how public budgets, incl. cohesion policy funds, and risk-mitigation instruments (e.g. guarantees, credit enhancement instruments) can be used strategically to improve the risk-return profile of investments that can attract commercial finance.

In Bulgaria, for examples, domestic financing (from public budgets, service providers revenues and potentially commercial finance) needs to be mobilised to reach the level of investment required to achieve compliance. This includes continuation of the water pricing reform, combined with targeted social measures to address affordability constraints and solidarity mechanisms to help cover investment costs in communities where financing capacities are especially limited. Authorities should review existing provisions that allow for service operators profits to accrue to government budgets rather than Water Associations, which could use such resources to strengthen the financial sustainability of the sector and provide a basis for accessing commercial finance. Building on recent set up of a Fund of Funds, in co-operation with European Bank for Reconstruction and Development (EBRD), options to attract commercial capital for creditworthy borrowers to finance water-related investments should be further explored.

Similar developments would benefit other countries. In Lithuania, existing financing structures such as the VIPA (Lithuanian **Public Investment Development Agency**) public fund managed by the Central Bank of Lithuania have a role to play. VIPA developed a dedicated financing instruments (the Water supply and wastewater fund and repayable assistance for the development of new wastewater networks). The fund could seek to attract private investment and facilitate the matching between the interests and needs of providers of private financing and project promoters. At a country workshop in October 2018, there were strong signals from International Financial Institutions that Romania could attract different kinds of support and funding as long as it is clearer in its strategy. This would also incentivise the private sector to invest in the Romanian water sector.

Tapping the potential of blended finance for water-related investments

The OECD defines blended finance as the strategic use of development finance (such as development assistance from donor governments and funds provided by philanthropic foundations), to mobilise additional commercial finance – (from public or private sources) –for investments that address the SDGs in developing countries.

The logic behind the approach is simple. Commercial investors, whether banks, investors, businesses or project developers respond to and are constrained by risks and returns associated with investments. As a result, investments with important public good dimensions may be backed by a sound business case but cannot necessarily be financed by commercial investors due to high risks associated with projects or uncertainty related to returns. In these cases, public support can be used strategically through blended finance to improve the 'risk-return' profile of investments, effectively de-risking investments to borrowers to access commercial finance. Blended approaches have a dual aim: i) mobilise additional capital for investments, and ii) serve a market building role, to help strengthen the financing systems upon which investment rely through greater accountability.

While the concept was developed with a focus on developing countries, the logic applies in European countries as well, in particular for water-related investments, where the public good dimension and political interference make financial returns uncertain.

The OECD has explored how blended finance could be accelerated for water-related investments, with a specific focus on water supply and sanitation utilities, off-grid sanitation and multi-purpose water infrastructure (see OECD, 2019b). Main lessons from this research are highlighted below and adjusted to the European context. Lessons from innovative arrangements in Europe could inspire further developments (see the Box 5.3 below).

Box 5.3. European experience with bending sources of finance for water-related investment

Although they do not all fit "neatly" into a blended finance definition, three recent developments in Europe can provide examples of mechanisms to leverage public/ development funds to access commercial capital.

NWB Bank in the Netherlands. Originally capitalised by public funds (Water boards [Regional water authorities] and government), the Bank now raises funds on capital markets to lend specifically for water-related projects (to the public sector – e.g. RWAs and other government authorities implementing projects). This is a fairly unique institution in the European context, but a good example other countries can learn from. For a concise summary, consult OECD (2014).

EIB's Sustainability Awareness Bond (SAB). The Bond was recently launched to raise debt financing focused in particular on water and wastewater projects. In a recent report, the OECD notes that "the new bond product aims at supporting the global goals by contributing to the development of a sustainable financial system through the financing of water and wastewater projects. In September 2018, the EIB issued its first EUR 500 million SAB. SAB eligibility is open to projects that contribute to the implementation of the SDGs without any geographical restriction. By December 2018, EUR 128 million has been allocated across 15 projects in 12 countries, including Senegal (28%), Italy (22%), Egypt (16%), and Panama (10%). Of these, 52% went to wastewater collection and treatment projects (EUR 65.6 million), 45% to water supply (EUR 55.8 million) and the remaining 3% to flood protection (EUR 3.6 million). SABs are a "use of proceeds" type of bond, meaning that disbursements for an SAB-eligible project will be funded from a dedicated account, where all funds raised through the issuance of SABs are managed by the EIB to finance solely water-related projects that meet the bond criteria." (OECD, 2019b).

As noted above, Bulgaria set up a Fund of Funds using a blended finance rationale and approach. The Fund of Funds will aim to use guarantees to help mobilise additional sources of capital. Further efforts could be developed to explore how public and development finance and risk-mitigation instruments (e.g. guarantees, credit enhancement instruments) can be used strategically to improve the risk-return profile of investments to attract commercial finance.

Source: OECD (2014), Water Governance in the Netherlands, OECD Publishing, Paris; OECD (2019b), Making Blended Finance work for water, OECD Publishing, Paris

A range of instruments are being used for blending, going beyond the more traditional loans and grants to the use of guarantees, securitisation, currency hedging, political risk insurance, etc. In this context, greater diversification of instruments could support better targeting of different risks and result in more commercial resources being targeted towards sustainable development outcomes. Amongst the different models, collective vehicles, such as funds, bring investors together to pool financing and offer opportunities for scaling up blended finance. In particular, structured funds allow donor governments to use concessional finance in a first loss position to provide a risk cushion for commercial investors. Blending can also occur through equity or debt investments in projects and companies in developing countries.

Beyond guarantees, technical assistance at the transaction level plays a major role in water and sanitation. Technical assistance can have different entry points in blended finance transactions, including for project development, investees such as utilities, or financiers such as banks to set up new lending programmes for the water and sanitation sector. Technical assistance has a particularly crucial role to play in tailoring existing blended finance structures to local contexts.

The success of blended finance is dependent on the ability to mobilise local commercial investment. Blended finance for water-related investments reinforces the need for, and benefits from, tailoring blended finance to the local context. Water and sanitation services are, by definition, locally sourced and provided and flood risks are best managed at the basin scale. At the same time, the sector requires strong public regulation due to the public good dimension of water and sanitation services and the common pool nature of water resources. These characteristics emphasise the need to work closely with local actors and align with local development needs.

There is a need to link blended finance approaches to the underlying value chain. To effectively tailor blended finance models for water-related investments, an understanding of the underlying business models and value chains is needed. Blended finance models can enter the sector at different points, for example at the water provision or treatment level, downstream at the end-user level or at the investor level. Effective blended finance approaches take into account the business models and respective revenue streams, and incorporate different stakeholder perspectives.

Pooling projects could be an effective way forward to address selected unfavourable project attributes. Providing commercial investors access to a variety of different transactions in the water and sanitation sector can mitigate concerns around small ticket size, risk exposure, limited sector or regional knowledge as well as high transaction costs. Pooling mechanisms such as blended finance funds tailor different risk and return profiles for individual investors, with development financiers often taking first loss and junior traches buffering the risk for commercial investors in the senior tranches. Guarantees, moreover, can strategically mitigate portfolio risk.

Beyond transaction-related insights into potential pathways to scale blended finance for water and sanitation, policy level implications can facilitate an uptake of blended solutions for sustainable development in the sector.

- Design blended finance in conjunction with efforts to improve the enabling environment. Blended finance cannot compensate for an unfavourable enabling environment, but rather needs to be accompanied by efforts to promote a stable and conducive policy environment. A weak enabling environment characterised by poorly designed or absent regulation, policies settings (e.g. water prices and tariffs), or institutional arrangements, compounded by political interference in the management of (often public) utilities, constrains commercial investment.

- Increase transparency to make a valid business case for commercial investment. Commercial investors are cautious about uncertainty regarding any of the risks related to an investment opportunity. With adequate contractual arrangement or blended instruments and mechanisms, it is possible to mitigate a variety of risks, share the remainder with the public sector or commercial co-investors, or take a certain level of risk on the financier's own book. However, in order to make such an assessment, risks associated with an investment should be transparent and hence quantifiable.

- Establish policy-level co-ordination and co-operation processes for blended finance. Coordination and co-operation among development finance actors on their blended finance engagements is a key for the market building aspect of blended finance, particularly when a concessional element is involved. Development financiers should co-ordinate more structurally beyond single transactions. Notably, an excessive reliance on concessional finance can inadvertently crowd out commercial finance, creating market distortions that impede greater accountability and financial sustainability of the sector. While there is general agreement about the need for improved co-operation, actions on the ground may remain fragmented.

In a longer term and dynamic perspective, blended finance is a transitory market building tool that is designed to enable stand-alone commercial investment in the long-run, by providing confidence, capacities and track record in markets where commercial investors are not yet present. Blended finance, starting with concessional elements, should phase out over time and ultimately exit in order to prevent market distortion. An analysis of the exit strategy should be integrated in any programme design.

5.3. Financing as part of flood risk mitigation strategies

Today, financing for flood risk management is employed in two directions (OECD, 2016):

- Investments in lowering flood risk – and thus investments in flood protection infrastructures, be it traditional "grey" infrastructures or nature-based solution (also known as pre-disaster options); and
- Provision of financial protection in case of flood events, thus refunding flood losses and damages (this financial protection is provided post-disaster, but it is arranged in a pre-disaster phase).

To date, flood protection in Europe is largely financed through public grants (Colgan et al, 2017). This can create significant costs for governments in terms of both investments in risk reduction and emergency responses and reconstruction (OECD, 2016). This can be especially burdensome in these times of growing public budget constraints.

Nevertheless, alternative instruments are available to finance both investments in flood protection and the provision of financial protection in case of flood events – and two categories of instruments can be identified, respectively: (Koehler et al, 2014):

- Economic Instruments (EIs), to provide a monetary/economic incentive promoting efficient flood risk management and risk reduction; they can be administered either by the government or by private entities. This category includes natural resource pricing, taxes, subsidies, marketable permits, payments for ecosystem services, licenses, property rights, habitat banking and trust funds;
- Risk Financing Instruments (RFI), comprising all instruments that promote the sharing and transfer of risks and losses. These are pre-disaster arrangements coming into play in a post-disaster phase. They include insurance, weather derivatives and catastrophe bonds.

These economic instruments, if properly designed, can contribute to achieve two key objectives of risk mitigation strategies (Koehler et al, 2014): raise revenues to finance flood protection (some instruments are not designed to raise revenues, but their application might result in resource saving); and indirectly incentivise behaviour and increase the uptake and efficiency of risk-reduction measures.

This suggests that such instruments should not only be looked at as financing sources/ mechanisms, but they should rather be intended as actual policy instruments to manage risk mitigation, together with other risk mitigation measures such as regulatory and research and development measures. They could be part of an integrated approach to risk mitigation strategies.

In the context of an integrated flood risk mitigation strategy, both EIs and RFIs can further contribute to reducing exposure, and in particular:

- EIs: as mentioned above, most of these mechanisms can raise new revenues to be invested in flood protection; often, this releases some of the existing pressure on public budgets. In addition, some EIs can influence behaviour, further reducing exposure levels – for example, by providing incentives for building and buying properties outside of at-risk areas;
- RFIs: insurance scheme, if properly designed, could steer behaviour towards reduced exposure levels. For example, a mandatory insurance can be required for buildings in at-risk areas, making it less convenient to live and install businesses in such areas.

5.3.1. Investments in lowering flood risk in Europe

Grants

Financial arrangements for investing in flood protection infrastructures vary from country to country, but usually there are under the responsibility of national governments and the EU.

European funds employed so far to finance flood protection (incl. infrastructure, nature-based solutions and other measures) include cohesion policy funds and European Investment Bank (EIB) funding. Nature-based solutions can also be funded through the LIFE programme – in addition, under this programme, the EIB administers the Natural Capital Financing Facility, which is designed to provide a "pipeline of bankable projects" involving natural capital, including natural infrastructure as adaptation to climate change.

In contrast, private conservations investments so far have not been so much directed at nature-based flood protection infrastructures, but rather at preserving and/or restoring natural systems – with flood reduction as a co-benefit only (Colgan et al, 2017).

Disaster relief is also provided by the EU Solidarity Fund, which was created as a reaction to the severe floods in Central Europe in the summer of 2002 and has been used for a range of catastrophic events[6].

Economic Instruments

EIs currently in place in Europe mainly belong to three categories:

- **Land use taxes and fees**: The flood-related land tax (or a flood-related component) can be applied to coastal or river flood risk management, to internalize the adverse effects of developments within high-risk areas. Land use taxes can represent a payment either for the land ownership itself or for its kind of use. Taxes and fees are also an important revenue-raising instrument, to raise funding for investments in flood protection infrastructures.
- **Subsidies**: Many forms of subsidies can intervene in managing flood risk. Subsidies can be on land prices in safe areas: while the demand for land increases in the subsidized safe areas, the demand for risky locations goes down decreasing land prices and pressure to develop there, and consequently decreasing costs for flood protection in the long term. Subsidies can also come along in the form of tax reduction – thus not necessarily as money transfers. Furthermore, some subsidies can have an adverse effect on flood risk levels.
- **Payment for Ecosystem Services** (PES) and Payments for Watershed Services (PWS): PES are voluntary mechanisms where suppliers of ecosystem goods and services (EGS) are paid by the beneficiaries to manage the ecosystems so that the provision of EGS is maintained and/or enhanced. PWS, in particular, are focused on the EGS provided by sound watershed management, linking upstream land and water management and downstream benefits. PWS are particularly relevant for financing natural water retention measures (NWRMs), as these measures are applied (or should be applied, to maximize their effectiveness) at the watershed level, and their impacts and related benefits also concern the watershed and, in particular, downstream areas.

Some of the existing EIs in Europe are presented in the Table 5.5 below[7].

Table 5.5. Selected existing economic instruments in Europe

Taxes for dykes maintenance (The Netherlands)	Historically, Dutch Water Boards (regional government bodies) have collected taxes to cover the costs for the maintenance of dykes. At present, Water Boards finance their water quality and flood protection activities (up to 95% of total costs) through local taxes. These taxes are charged according to the `beneficiary-pays' principle: the beneficiaries pay a water board tax proportionally to their interests. Thus different stakeholders (farmers, residents, industries) may be charged different taxes by the same water board. (Source: Filatova, 2014)
Taxes for landowners protected by a dyke (Germany)	Wadden Sea States are responsible for the organisation and administration of (public) coastal defence in Germany. However, as coastal defence has national consequences, capital measures are co-financed by the federal government with 70% of total eligible costs (the other 30% are matched by the states). The maintenance of existing state coastal defence structures is financed 100% by the state. Municipalities and/or local water boards that are responsible for coastal defence measures in their area normally have to contribute between 5 and 20%. A small but increasing contribution to coastal defence comes from the European Union. The overall principle is that all persons who profit from protection are in charge of maintaining the dikes. Beneficiaries are organised in water boards which have to do the maintenance and construction works on the mainland dikes, except for some which are under State responsibility. (Sources: Trilateral Working Group on Coastal Protection and Sea Level Rise - CPSL, 2010; CPSL, 2001)
Drinking water forests – Offset scheme1 (Germany)	Bionade Corporation, a beverage manufacturer, in partnership with Trinkwasserwald e.V. (Drinking Water Forest Association), is creating '"drinking water forests" all over Germany. The project involves afforestation of privately and publicly owned land with deciduous broadleaved trees, with the aim of enhancing groundwater regeneration. Bionade aims to offset its own water use in doing so, with a target of about 100 million liters each year or 130 hectares of reforested lands. Forest land holders sign contracts with Trinkwasserwald e.V. agreeing to reforest monoculture coniferous plots with deciduous trees.
Bosco Limite project – PES scheme2 (Italy)	The project aims at creating a forested area that will catch precipitation and increase groundwater recharge and other EGS. The project was implemented in an area previously used for intensive agriculture, and activities with high economic returns. In turn, it provides in fact a wide range of goods and services, such as groundwater recharge, runoff reduction (and thus flood mitigation), CO_2 fixation, biodiversity safeguard, production of high quality wood and biomass for energy production, and recreational-touristic services. Such services have then become alternative, competitive sources of income for landowners who made their land available for reforestation, through the establishment of a PES scheme.

Notes:
1. NWRM project - http://nwrm.eu/sites/default/files/sd11_final_version.pdf
2. NWRM project - http://nwrm.eu/sites/default/files/sd11_final_version.pdf

Other instruments

A particular financing instrument, which cannot be defined as a EI, is the Barnier Fund in France. In essence, the Fund is mostly financed by a 12% levy on the compulsory insurance against natural disasters for residential and commercial/industrial buildings. The State's contribution to the Fund is around 7% of total revenues, whereas revenues from the levy cover up to 97% of the total expenditures of the Fund. The Fund was set up to finance some flood protection actions and it is almost self- sustainable. A detailed analysis of Barnier Fund is appended.

Green bonds can also be used to fund flood protection infrastructures. "Green bonds" is the term that is often applied to environmentally related impact investing. The largest type of green bonds is funding related to climate change, predominantly for projects designed to reduce greenhouse gas emissions. To access the green bond market, nature-based infrastructure projects need to meet two conditions: i) a revenue source to repay the bond buyers; and ii) a set of performance standards to demonstrate attainment of flood risk reduction goals. Repayment of the bonds can come either from public funding or from private funding, in the form of private organisation created to share benefits among members/funders. Green bonds can be financed both by public or private investors. In the US (and, to a lesser extent, in Europe), the revenue flow stream for private investment in green bonds is often provided through "special purpose districts". These districts manage "semi-public" infrastructure, and there is a variety of options for the structure, financing, and governance of such districts. Revenues are generated as these districts directly link the beneficiaries of infrastructures to the financing of infrastructures, so they can back green bonds (Colgan et al, 2017).

In the UK, the National Infrastructure Commission reported that many UK pension funds and other institutional investors have developed investment funds or investment platforms for infrastructures.

5.3.2. Provision of financial protection – Risk Financing Instruments in Europe

In some countries **Risk-Financing Instruments** (RFIs) are in place: this category comprises all instruments that promote the sharing and transfer of risks and losses. These are pre-disaster arrangements coming into play in a post-disaster phase. They include insurance, green bonds, weather derivatives and catastrophe bonds.

In Europe and elsewhere, the most common RFI is insurance, an instrument providing financial protection against flood damages. It aims at eliminating economically-unwarranted use of flood-prone area, while not prohibiting land use. Insurance serves as a risk-sharing and risk-communication device to help individuals rationalise their land use choices in at-risk areas. If risks are correctly priced in premiums, insurance allows location in hazard-prone areas for those who are ready to bear the risk, without increasing a burden on taxpayers (Filatova, 2014). In fact, the level of insurance penetration has been shown to be negatively correlated with the level of impact of disasters on economic output – in other words, countries with higher levels of insurance penetration face more limited negative impacts on economic output (OECD, 2016).

Nevertheless, the way insurance is designed and premiums are priced is key to the good functioning of insurances as a risk-communication instrument. In particular (Filatova, 2014), insurance premiums that are uniform or do not differentiate for actual flood risk may bias economic location decisions; non-compulsory flood insurance leads to information asymmetry among residential buyers and sellers.

Table 5.6 below provides an overview of existing flood insurance schemes in the EU.

Table 5.6. Selected existing insurance schemes in the EU

Mandatory multi-risk home insurance (Romania)	In Romania, homeowners are required to purchase home insurance covering damages from floods, landslides and earthquakes –they can be fined if not insured. Nevertheless, many homeowners do not purchase any insurance, because some legal clauses allow them to remain uninsured if some socio-economic thresholds are not met. As a result, only 38% of dwellings are currently covered by insurance. Overall, it was observed that the Romanian disaster risk financing framework, in its current form, is rather ex-post, and the link between risk reduction and risk financing is weak (Surminski and Hudson, 2017).
Flood insurance in the Po RBD (Italy)	In the Po RBD, flood risk management involves the controlled flooding of agricultural land (low-value use) in order to avoid larger losses in urban areas (high-value use). However, in agricultural areas, only some 5% of private properties at-risk are covered by flood insurance. In other words, flood insurance here is specifically designed to cover the deliberate costs arising from a risk reduction scheme (i.e. the temporary flooding of agricultural land to preserve urban areas) (Surminski and Hudson, 2017).
Flood Re scheme (the UK)	Flood Re scheme is a re-insurance scheme negotiated through a series of voluntary agreements between the Government and members of the Association of British Insurers (ABI). It has been set up to help those households who live in a flood risk area find affordable home insurance[1]. The aim is to ensure the availability and affordability of flood insurance, without placing unsustainable costs on wider policyholders and taxpayers. The scheme does not contain any risk-reduction element, although the Government is showing some commitment to improve flood risk management (Surminski and Hudson, 2017).
CatNat - Natural disaster insurance system (France)	CatNat is a public-private compensation system that covers losses that cannot be insured in private markets, such as flooding. Under CatNat, it is mandatory for insurers to extend property and vehicle insurance contracts to cover damage caused by natural disasters. These additional premiums are not differentiated according to the actual natural disaster risk, but are fixed by the Government following a principle of national solidarity. One of the main reinsurers providing coverage for CatNat is the Central Fund for Reinsurance[2] (CCR), an international reinsurance company owned by the French government. The government will compensate damage above a certain amount stipulated in the law by providing an unlimited guarantee of compensation exclusively to the CCR, and not to other reinsurers in the market. The scheme is linked to Natural Risk Prevention Plans, which regulate land use to reduce exposure of property and people to natural hazards – mostly by limiting new construction and enforcing implementation of prevention measures by local communities in flood-prone areas (Poussin et al, 2013).

Note: 1. https://www.floodre.co.uk/

5.3.3. Innovative financing mechanisms from outside the EU

Innovative financing mechanisms have the potential to expand available funding options for flood protection. Some of these instruments are already used outside of Europe. Innovative mechanisms all belong the EIs category. See below for illustrations.

Table 5.7. Innovative financing mechanisms for flood protection

Instrument	Description	Examples from outside the EU
Marketable permits	A scheme was proposed in which government issues the socially-optimal amount of marketable permits for developing risk zones. The amount of marketable permits should differ per zone depending on the probability of hazard occurrence or other criteria. The market further allocates available land to the highest opportunity cost.	No examples available: instrument proposed at the research/ policy level, and never implemented.
Advanced market commitment	The government guarantees a certain income to the entity providing a desired activity, making this instrument comparable to a subsidy.	This instrument has not been applied to managing flood risk yet –although there is a potential for its application.
Transferable development rights (TDR)	A cap is set on the quantity of development and the area is split into the receiving zone (with lower or no flood risk) and sending zone (with higher risk). Landowners in the sending zone cannot use their development rights, but can sell them to someone in the receiving zone. Thus, TDR discourage development in at-risk areas and move it to low-risk areas, or areas which are easier to protect against floods. Thus, TDR are not a direct financing scheme for flood protection, but save public resource in many ways: i) it moves development in areas with lower flood protection costs; ii) it reduces transaction and administrative costs, as compared to the management of a traditional permit system; and iii) it reduces the costs of damages in case of flood events.	TDRs are common in the US for nature conservation projects – in fact, TDR schemes to manage retreat from flood-prone areas can go hand in hand with nature conservation initiatives (e.g. nature based solutions for flood risk management).
Water Funds (and Trust Funds in general)	Water funds finance watershed management, paying for the services that ecosystems provide to humans. Water funds pool together capital contributions from different stakeholders involved in watershed management such as water supply companies, hydropower plants, irrigation districts and agricultural associations. Capital contributions are invested in the financial market through trust funds, and the financial returns are invested in watershed management activities, such as conservation measures, protected areas, promotion of eco-friendly agriculture and so on. Trust funds have been used widely to finance nature conservation worldwide and in particular in Latin America and Asia.	Their implementation has been spreading in recent years and in Latin America in particular. The FONAG (Fondo para la Protección del Agua) is an example of this.
Habitat banking	Habitat banking aims at conserving the ecosystem services of land, including biodiversity. Credits are given for the creation, restoration and enhancement of habitats, while debits occur when ecosystems are unavoidably degraded or destroyed.	No examples available to the authors' knowledge

Source: Koehler et al, 2014; Filatova, 2014, and NWRM Project.[8]

The following criteria were used to assess the potential of existing instruments as part of risk mitigation strategies:

- What is the revenue-raising capacity of the instrument?
- What is the capacity of the instrument to steer behaviour and minimize flood risk?
- Is the financing source adaptable? In other words: is the instrument apt to finance large, on-time investments (e.g. in large infrastructures), or rather modular, smaller investments over a period of time?
- Does the instrument decrease reliance on public budgets? Or, in other words: how does it allocate the costs to investors other than public actors?
- What is the geographical scale of the instrument? And what is its potential to be scaled up? Is it replicable in other countries? Which requisites must be met?

Table 5.8 below summarises the assessment of the relevance of innovative instruments as part of integrated flood protection strategies in Europe. A more detailed analysis of EIs and RFIs is appended.

Table 5.8. The performance of economic policy instruments as part of risk mitigation strategies

Performance criteria / Economic instrument	1. Revenue-raising capacity	2. Capacity to steer behaviour	3. Adaptability of the source	4. Allocation of costs across actors	5. Geographical scale	6. Replicability
Land taxes	Can be significant	Yes, if proper rates	Yes (possible to plan)	Individuals in at risk-areas pay, but still some burden	All scales	Yes
Earmarking water charges	Not possible to estimate	No	Yes (possible to plan)	Water users pay, but still some burden	All scales	Yes
Offset schemes	Yes, but only as in integration to other sources	Not applicable	Low - focus on specific protection or restoration actions	Yes, private actors pays, but usually for specific measures	Localized measures and interventions	Yes, in principle
PES schemes	Yes, but only as in integration to other sources	Yes, but the scale depend on the scale of the scheme	Low - focus on specific measures or practices	Yes, private actors pays, but usually for specific measures or services	Often local or watershed scale	Yes, but there might be constraints
Green bonds	Significant capacity	It depends on the way revenues are raised	Mostly used for infrastructure development	Opportunity to boost private investment	All scales	To be investigated (for the EU)
Flood insurance schemes	Not applicable	Depending on how they are designed – Not at the moment	Not applicable	If properly designed, they can reduce burden on the long run	National level	Yes
Fonds Barnier	Yes – main source of disaster prevention	Not really, although it could	Wide range of expenditures	All building owners are charges, releasing burden	National scale	Countries where flood insurance is mandatory

Note: Colours reflect a traffic light approach. Grey stands for "not applicable"

In summary:

- **Land taxes** have a good revenue-raising potential, as well as a good potential to steer behaviour, they can be applied at all scales and in all countries;
- **Earmarking water charges** have a good revenue-raising potential and can be applied everywhere at all scales; nevertheless, they charge water user for flood protection, thus providing no incentive at all for risk-reduction behaviour (basically, the wrong people are charged!);
- **Offset schemes** offer the possibility to introduce private capital into nature protection and restoration, but they normally focus on specific measures and/or actions;
- **PES schemes** introduce some private financing for environmentally-sound management and nature protection, but usually for specific measures or services;
- **Green bonds** have a significant revenue-raising capacity, and offer the opportunity to inject substantial private investment into flood protection actions. However, their uptake is low in Europe, and further investigation is needed to understand why;
- **Flood insurance schemes**, if properly designed, could provide a strong incentive for risk-reduction behaviour – thus also reducing the need for public investment on the long run. However, existing schemes do not provide such incentives;

- The **Fonds Barnier** is a mechanism able at financing most of flood prevention actions in France. However, it does not really provide an incentive for risk-reduction behaviour, and a similar mechanism could only be implemented in countries with compulsory flood insurance for buildings.

Different instruments can be combined in risk mitigation strategies, to get the most out of each of them, and the best mix of instruments will need to be assessed on a case-by-case basis.

References

Ambec, S. et al. (2016), *Review on international best practice on charges for water management*, Toulouse School of Economics, OECD Background Paper (unpublished).

Ashley R., Horton B., Boxall J., Speight V. (2018), *Financing Water in 28 European Countries. Baseline Report*, Background Paper (unpublished)

Biggs, C. et al. (2009), *Distributed Water Systems: A networked and localised approach for sustainable water services. Business Intelligence and Policy Instruments*, Victorian Eco-innovation Lab, University of Melbourne, www.ecoinnovationlab.com/website/wp-content/attachments/234_Distributed-Water-Systems.VEIL.pdf .

Borsje B., de Vries S., Janssen S.K.H., Luijendijk A.P., Vuik V. (2017), Building with nature as coastal protection strategy in the Netherlands. In: *Living shorelines. The science and management of nature-based coastal protection*. Editors: Bilkovic D.M., Mitchell M.M., La Peyre M.K., Toft J.D. CRC press.

Broeckx S., Smets S., Liekens I., Bulckaen D., Nocker L. (2011), Designing a long-term flood risk management plan for the Scheldt estuary using a risk-based approach. *Natural Hazards* vol. 57, issue 2, pp. 245-266.

Brugge, R. (van der), R. Graaf (de) (2010), Linking water policy innovation and urban renewal: the case of Rotterdam, the Netherlands, *Water Policy*, Vol. 12, pp. 381-400.

Bullock J.M., Aronson J., Newton A.C., Pywell R.F., Rey-Benayas J.M. (2011), Restoration of ecosystem services and biodiversity: conflicts and opportunities. *Trends in Ecology & Evolution* 26(10): 541-549.

CEREMA (2014), *Coût des protections contre les inondations fluviales*, CEREMA, Lyons, France.

Colgan C.S., Beck M.W., Narayan S. (2017), *Financing Natural Infrastructure for Coastal Flood Damage Reduction*. Lloyd's Tercentenary Research Foundation, London. https://conservationgateway.org/ConservationPractices/Marine/crr/library/Documents/FinancingNaturalInfrastructureReport.pdf

CPB (2000), *Ruimte voor water: Kosten en baten van zes projecten en enige alternatieven*, Working Document 130, CPB, The Hague, Netherlands.

CPSL (2010), *Third Report, the Role of Spatial Planning and Sediment in Coastal Risk Management*, Wadden Sea Ecosystem No. 28., Common Wadden Sea Secretariat, Wilhelmshaven, Germany

CPSL (2001), *Final Report of the Trilateral Working Group on Coastal Protection and Sea Level Rise*, Wadden Sea Ecosystem No. 13, Common Wadden Sea Secretariat, Wilhelmshaven, Germany

European Environment Agency (2017), *GI and Flood Management: Promoting cost-efficient flood risk reduction via GI solutions*, EEA Report, No 14/2017, EEA, Copenhagen, Denmark.

Environment Agency (UK, 2017), *Working with Natural Processes: Evidence Directory*. https://www.gov.uk/government/publications/working-with-natural-processes-to-reduce-flood-risk

European Commission (2013), *Building a Green Infrastructure for Europe*, Brussels, doi: 10.2779/54125

Feyen L., Dankers R., Bodis K., Salamon P., Barredo J.I. (2012), Fluvial flood risk in Europe in present and future climates. *Climatic Change* vol. 112, issue 1, pp 47-62.

Filatova T. (2014), Market-based instruments for flood risk management: A review of theory, practice and perspectives for climate adaptation policy. *Environmental Science and Policy* 37, pp 227-242.

FLOODsite Project Deliverable D9.1 (2007), *Evaluating flood damages: guidance and recommendations on principles and methods*. Report number T09-06-01.

Genovese E., Lugeri N., Lavalle C., Barredo J.I., Bindi M., Moriondo M. (2007), *An assessment of weather related risks in Europe: maps of flood and drought risk*. Interim report for ADAM FP6 Integrated Project, JRC Scientific and Technical Reports, ISSN 1018-5593.

Gleick, P. (2002), Water management: Soft water paths, *Nature*, Vol. 418, p. 373 (25 July 2002), doi:10.1038/418373a.

Grossmann, M., Hartje, V. and Meyerhoff, J. (2010), *Ökonomische Bewertung naturverträglicher Hochwasservorsorge an der Elbe — Naturschutz und Biologische Vielfalt*, Federal Agency for Nature Conservation, Bonn, Germany.

ICPR (2006), *Nackweis der Wirksamkeit von Massnahmen zur Minderung der Hochwasserstaende im Rhein infolge Umsetzung des Aktionsplans Hochwasser bis 2005*. Report number 157. International Commission for the Protection of the Rhine, Koblenz.

Kind J.M. (2014), Economically efficient flood protections standards for the Netherlands. *Journal of Flood Risk Management* 7, pp 103-117.

Kind J.M., Botzen W.J., Aerts, C.J.H. (2017), Accounting for risk aversion, income distribution and social welfare in cost-benefit analysis for flood risk management. Clim*ate Change* 8.

Koehler M., Mechler R., Botzen W., Surminski S., Pulido Velázquez M., Leblois A., Keating A., Mochizuki J., Manez M., Cremades R., Hall J. (2014), *Review of economic instruments in risk reduction*. ENHANCE (Enhancing Risk Management Partnerships for Catastrophic Natural Disasters in Europe) Research project, Deliverable 5.1, April 2014.
http://enhanceproject.eu/uploads/deliverable/file/21/ENHANCE_D5.1_Review_of_economic_instruments_in_risk_reduction.pdf

Lugeri N, Kundzewicz Z.W., Genovese E., Hochrainer W., Radziejewski M. (2010), River flood risk and adaptation in Europe – assessment of the present status. *Mitigation and Adaptation Strategies for Global Change*, 15(7):621-639, DOI: 10.1007/s11027-009-9211-8

Mazza L., Bennet G., De knocker L., Gantioler S., Losarcos L., Margerison C., Kaphengst T., McConville A., Rayment M., ten Brink P., Tucker G., van Diggelen R. (2011), *Green inrfrastructure implementation and efficiency: Freshwater and wetlands management and restoration*. Published by IEEP.

Naumann S., Davis M., Kaphengst T., Pieterse M., Rayment M. (2011), *Design, implementation and cost elements of GI projects: Final report to the European Commission*, Directorate-General of the Environment, Ecologic and GHK.

NWRM (2016), *53 National Water Retention Measures illustrated*.
http://nwrm.eu/sites/default/files/documents-docs/53-nwrm-illustrated.pdf

OECD (2020), *Addressing the Social Consequences of Water Tariffs*, Working Paper, OECD

OECD (2019a), *Land Value Capture: Framework and Instruments*, unpublished paper

OECD (2019b), *Making Blended Finance Work for Water and Sanitation: Unlocking Commercial Finance for SDG 6*, OECD Studies on Water, OECD Publishing, Paris, https://doi.org/10.1787/5efc8950-en

OECD (2017), *Groundwater Allocation: Managing Growing Pressures on Quantity and Quality*, OECD Publishing, Paris, https://doi.org/10.1787/9789264281554-en

OECD (2016), *Financial management of flood risk*, OECD Publishing, Paris, http://www.oecd.org/finance/financial-management-of-flood-risk.htm

OECD (2015a), *The Governance of Water Regulators*, OECD Studies on Water, OECD Publishing, Paris, https://dx.doi.org/10.1787/9789264231092-en

OECD (2015b), *Water Resources Allocation. Sharing Risks and Opportunities*, OECD Studies on Water, OECD Publishing, Paris, https://dx.doi.org/10.1787/9789264229631-en

OECD (2015c), *Water and Cities*, OECD Studies on Water, OECD Publishing, Paris, https://dx.doi.org/10.1787/9789264230149-en

OECD (2014), *Water Governance in the Netherlands. Fit for the Future?*, OECD Studies on Water, OECD Publishing, Paris, http://dx.doi.org/10.1787/9789264102637-en

OECD (2013a), *Water and Climate Change Adaptation: Policies to Navigate Uncharted Waters*, OECD

Studies on Water, OECD Publishing, Paris, https://dx.doi.org/10.1787/9789264200449-en

OECD (2013b), *Water Security for Better Lives*, OECD Studies on Water, OECD Publishing, Paris, http://dx.doi.org/10.1787/9789264202405-en.

Poussin J.K., Botzen W.J.W., Aerts J.C. (2013), Stimulating flood damage mitigation through insurance: an assessment of the French CatNat system, *Environmental Hazards*, 12:3-4, 258-277, DOI: 10.1080/17477891.2013.832650

Pugh T.A.M, Mackenzie A.R., Whyatt D., Hewitt C.N. (2012), Effectiveness of Green Infrastructure for Improvement of Air Quality in Urban Street Canyons, *Environmental Science & Technology*, 46(14):7692-9, DOI: 10.1021/es300826w

Scussolini P., Aerts J.C.J.H., Jongman B., Bouwer L.M., Winsemius H.C., de Moel H., Ward P.J. (2016), FLOPROS: an evolving global database of flood protection standards.

Surminski S., Hudson P. (2017), Investigating the Risk Reduction Potential of Disaster Insurance Across Europe, *The Geneva Papers on Risk and Insurance - Issues and Practice*, 2017, vol. 42, issue 2, 247-274

Tröltzsch, J., Görlach, B., Lückge, H., Peter, H., Sartorius, C. (2012), *Kosten und Nutzen von Anpassungsmaßnahmen an den Klimawandel: Analyse von 28 Anpassungsmaßnahmen in Deutschland*, Report for Umweltbundesamt, Ecologic Institute, INFRAS and Fraunhofer Institut.

UNEP (2014), *Green Infrastructure: Guide for Water Management*, http://hdl.handle.net/20.500.11822/9291

UNESCO (2017), *World Water Development Report: Wastewater: the Untapped Resource*, UNESCO, Paris, ISBN:978-92-3-100201-4

WGF (2012), *Flood risk Management, Economics and Decision Making Support*. Resource document of the Working Group Floods of the Common Implementation Strategy for the Water Framework Directive.

Schwarz U., Batrich C., Hulea O., Moroz S., Pumputyte N. Rast G., Bern M., Siposs V. (2006), *Floods in the Danube River Basin: Flood risk mitigation for people living along the Danube – The potential for floodplain protection and restoration*, WWF, Geneva.

Zhou N., Williams, C.J. (2013), *An International Review of Eco-city. Theory, Indicators and Case Studies*, LBNL 6513, Berkeley

Notes

[1] As noted in previous sections, the position of Finland may result from an underestimate of current levels of expenditures.

[2] Note that even relatively arid countries can face increasing flood risks. The Northern and Southern regions of Span are good examples. And Portugal is projected to face heavier rain.

[3] EU funding offers that kind of support.

[4] There have been a number of initiatives to try to account for the wider social benefits of NBS, which draw on the development of economic valuation techniques for ecosystem services concepts; social value; and natural capital accounting.

[5] The costs and benefits of both grey and GI measures presented in this study are not directly comparable as they are highly dependent on the geographical location of the measure.

[6] http://ec.europa.eu/regional_policy/en/funding/solidarity-fund/

[7] The literature on economic instruments for flood protection often quotes other instruments, such as for example catastrophe taxes, lower land taxes in areas with lower risk, water markets for indirect risk reduction. However, no more detailed information was found, so the table only includes those instruments for which a good level of information could be found.

[8] NWRM Project - Synthesis document No. 11: Financing of Natural Water Retention Measures. http://nwrm.eu/sites/default/files/sd11_final_version.pdf

Annex A. Costs of addressing emerging challenges in wastewater collection and treatment

The Table below summarises data collected in the literature, on the possible costs of emerging challenges for water service provision. The focus is on managing combined sewer overflows, on enhancing the treatment of waste water to address contaminants of emerging concern and other priority substances, and related issues.

Note that although there will be added costs for managing microplastics in the water services system, there is as yet scant information about this. The major costs are likely to be in the sludge handling, as the continuing spreading of sludge to land (preferred option in many member states) may be compromised if microplastics and other micropollutants are not controlled at source.

Table A A.1. Cost estimates for managing emerging and associated pollutants

Source and applicability	Relevance	Cost ranges
		CSO spills
USA – reduce/stop spills (Renn, 2016)	Costs for the entire USA for compliance with Clean Water Act	$48bn US capital remediation bill for CSOs. Not only for emerging pollutants.
EU (EC, 2011)	Relates to the operational costs of CSOs, not only for emerging pollutants	Additional costs for adding to the PS list €18 per cap per year. EurEau (2012) suggest this estimate is too low. Actual costs will be some 25-50% added on to original annual costs, prior to the revision.
Belgium (Dirckxe et al, 2011) (25,000pe)	Relates to the operational costs of CSOs, not only for emerging pollutants	€10m – 100m disconnection to reduce spills between 30 and 100% €4m – 10m for storage tanks to reduce spills by between 60 and 100% €100k – 900k throttle through flows reducing spills by 20 – 35% €100k – 700k RTC to reduce spills by 10 – 75%
Germany (Tondera et al, 2017)	Added costs for emerging pollutants	Costs for disinfection only: Capital costs €275/m^3 treated, operational cost €5-10/m^3.
EurEau (2018) Table 2.4. Spain.	For 10 years. Relates to the operational costs of CSOs, not only for emerging pollutants	€3 per cap/year (includes climate change)
		Wastewater Treatment
Finland (Katko, 2016)	Refers to all enhancements to treatment	Enhancing removal of nitrogen from 70 to 90% (450 WWTP) €400-600m. For phosphorus €200-600m. Recovered P value only €2-4m for further costs of €60-90m. Pharmaceuticals and hazardous substances removal €700-1400m.
Germany (Entec, 2011).	Both compliance costs for treatment and added costs for certain emerging pollutants	Overall compliance costs €5bn – 12bn, depending on treatment. Tertiary treatment €5-11m per WWTP with added energy and CO_2 costs. Consequent additional sludge disposal costs €86m – 256bn per year. Average NPV capital and operating costs are €398 per p.e. and €295 per p.e. for Diclofenac removal using either GAC (99% removal) or UV (57% removal). Case study for WWTP in Ulm, with 440,000 population, costs of €40m.
Spain (Entec, 2011)	Emerging pollutants deemed to not add to costs in this case	No significant additional costs as the PEC is lower than the specified EQS.
UK (Comber, 2007)	Additional costs for certain emerging substances	Added costs to tackle APIs for UK as a whole €10bn, or potentially some €1.5m per WWTP. This assumes all 6800 WWTP need to be upgraded. Added costs for pharmaceutical removal was some €0.8 – 25m capital and €0.02 –

Source and applicability	Relevance	Cost ranges
UK (EC, 2011)	Additional costs for certain emerging substances	4.1m operational costs per year. Cost estimates for E2 removal: €18 per cap per year.
UK (Entec, 2011)	Additional costs for certain emerging substances	Yorkshire Region 48 WWTP costs: €725m, with operational costs of €45m per year. Discounted costs are €1020m discounted over 25 years. Scaling these figures up to England and Wales: €12 - 14bn.
Switzerland (Entec, 2011)	Additional costs for certain emerging substances	756 WWTP in Switzerland. Diclofenac removal costs: €495 – 591m capital costs and €56-76m operational costs (p.e. ranges from 14,000 – 590,000) – total costs were some €0.03-0.3/year per m^3 treated. Overall
Switzerland (Beiber et al, 2018)	Additional costs for certain emerging substances	123 WWTP, serving >80,000 population, discharging to surface water and or into drinking water sources, need upgrading out of 756 total. Some 50% of Swiss effluent will be treated with 80% removal of micropollutants. Total costs are €1bn. Annual costs are €115m. Discounted total costs are €2.8bn. Charges capped at €8 per year/inhabitant.
Switzerland (EC, 2011)	Additional costs for certain emerging substances	Additional 5 – 25% on conventional treatment costs, or some €11-18 per inhabitant.
Cyprus (EurEau, 2018)	For 10 years Additional costs for certain emerging substances	€1 per cap/year (includes climate change)
Denmark (EurEau, 2018)	For 10 years Additional costs for certain emerging substances	€96 per cap/year (includes climate change)
Spain (EurEau, 2018)	For 10 years Additional costs for certain emerging substances	€3 per cap/year (includes climate change)
France (EurEau, 2018)	For 10 years Additional costs for certain emerging substances	€22 per cap/year (includes climate change)
Italy (EurEau, 2018) Table 2.4	For 10 years Additional costs for certain emerging substances	€17 per cap/year (includes climate change) and covers all water services, not only WW.
Ireland (EurEau, 2018) Table 2.4	For 10 years Additional costs for certain emerging substances	€17 per cap/year (includes climate change) and covers all water services, not only WW.
Norway (EurEau, 2018) Table 2.4	For 10 years Additional costs for certain emerging substances	€44 per cap/year (includes climate change) and covers all water services, not only WW.
Norway (EurEau, 2018) Table 2.4	For 10 years Additional costs for certain emerging substances	€44 per cap/year (includes climate change) and covers all water services, not only WW.
Netherlands (EurEau, 2018) Table 2.4	For 10 years Additional costs for certain emerging substances	€180 per cap/year (includes climate change) and covers all water services, not only WW.
	Other	
Spain (EurEau, 2018) Table 2.4	For 10 years Additional costs for certain emerging substances	€3 per cap/year (includes climate change) for drinking water remediation
Denmark (EurEau, 2018) Table 2.4	For 10 years Additional costs for certain emerging substances	€21 per cap /year (includes climate change) for drinking water remediation
France (EurEau, 2018) Table 2.4	For 10 years Additional costs for certain emerging substances	€0.1 per cap /year (includes climate change) for drinking water remediation
EU (EC, 2011)	Additional costs for certain emerging substances, but for added monitoring	The 20,900m^3 of drinking water abstracted for drinking water production require pesticide removal costing €0.028/m^3. Estimates indicate that some 74% of surface waters used for this need treatment. Monitoring costs across MS, for additional PHS and pharmaceuticals as some €15-36m per year, adding 22-52% to the original annual monitoring costs of €69m (51-97m) prior to the additions to PHS. Of this, added pharmaceutical monitoring costs are €3-6m per year.
UK (EC, 2011)	Nickel	Nickel is highest in UK waters. Removal costs are €2bn capital with added ongoing costs.

Source: Data compiled by Richard Ashley, for the OECD (Ashley R., Horton B., Boxall J., Speight V., 2018, Financing Water in 28 European Countries. Baseline Report, Background Paper (unpublished).

Annex B. Data and method

Current levels of expenditure on water supply and sanitation are based on Eurostat data, which combines a range of data sets covering water-related public and household expenditures.

Business-as-usual scenario projections on future expenditures for water supply and sanitation are driven by the growth in urban population. Additional scenarios were developed for water supply and sanitation to factor in such drivers as compliance with current and emerging regulation. The additional scenarios are based on available expenditure reported by countries and each countries distance to compliance with articles three, four, and five.

Projections on future expenditures for flood protection combine estimates of exposure (of population, assets and GDP) to risks of coastal or river floods.

The characterisation of past sources of financing in each country is derived from data on current levels of expenditures, as well as complementary data sources on debt finance and EU funding.

Countries' financing capacities are proxied by analysing latitude in 3 areas: ability to raise the price of water services (taking into account affordability concerns), ability to increase public spending (based on taxes or borrowing), and ability to tap into private finance.

A separate report presents the methods and data used in more details.

Annex C. Data supporting the results on projected coastal flood risk investment needs

Table A C.1. Data supporting the results on projected coastal flood risk investment needs

	Maritime basin	Expected sea level rise	Coastal length	Zone below 5m elevation	Coastline subject to erosion		GDP in 50 km zone		Population in 50 km zone		Change in built-up in areas vulnerable to coastal floods	People in the 100-year flood plain	People flooded	Damage costs	Exp. to protect against coastal flood risk
			km	%	km	%	million €	%	number	%	%-increase 2050 Brown et al., (2011)	Million 2030 Neumann et al, (2015)	Thousands/year 2050 Hinkel et al, (2010)	Billion Euro/year 2050 Hinkel et al, (2010)	Category 1-4
Austria	-	-	-	-	-	-	-	-	-	-	-	-	-	-	-
Belgium	North Sea	High	98	>85%	25	25	95.722	34	3.846.676	37	10,34	-	1,9	1,1	3
Bulgaria	Black Sea	Medium	125	<5%	26	45	2.776	5	39.064	5	0	-	0,2	<0,1	1
Croatia	-	-	-	-	-	-	-	-	-	-	-	-	-	-	-
Cyprus	Mediterranean Sea	Medium	367	<5%	110	30	2.907	100	730.367	100	60	-	0,1	<0,1	1
Czech Republic	-	-	-	-	-	-	-	-	-	-	-	-	-	-	-
Denmark	North Sea Baltic Sea	North Sea: High Baltic Sea: Low	4.605	22%	607	13	104.043	72	5.397.640	100	18,69	-	0,5	0,5	2
Estonia	Baltic Sea	Low	2.549	10-15%	51	2	10.646	66	959.259	71	0	-	0,1	<0,1	1
Finland	Baltic Sea	Low	14.018	<5%	5	0,04	65.201	50	2.975.247	57	2,17	-	0,3	0,2	1
France	North Sea Atlantic Ocean Mediterranean Sea	North Sea: High Atlantic Ocean: High Mediterranean Sea: Mediu	8.245	4,70%	2055	25	256.430	17	16.185.472	26	7,13	-	3,5	2,5	4

	Maritime basin	Expected sea level rise	Coastal length	Zone below 5m elevation	Coastline subject to erosion		GDP in 50 km zone		Population in 50 km zone	Change in built-up in areas vulnerable to coastal floods	People in the 100-year flood plain	People flooded	Damage costs	Exp. to protect against coastal flood risk
Germany	North Sea Baltic Sea	North Sea: High Baltic Sea: Low	3.524	82%	452	26	10.147	5	5.777.217	7	1,36	3,2	2	3	3
Greece	Mediterranean Sea	Medium	13.780	<5%	3.945	28,6	140268	69	10.157.398	92	3,57	-	0,5	<0,1	1
Hungary	-	-	-	-	-	-	-	-	-	-	-	-	-	-	-
Ireland	Atlantic Ocean	High	4.577	<5%	912	20	71.505	58	3.343.018	83	21,43	-	0,6	<0,1	1
Italy	Mediterranean Sea	Medium	7.468	<5%	1704	22,8	559.306	42	34.154.065	59	0	2,4	1,1	0,3	3
Latvia	Baltic Sea	Low	534	<5%	175	33	16.306	72	1.484.290	64	0	-	0,8	<0,1	1
Lithuania	Baltic Sea	Low	262	<5%	64	24	3126	8	423.503	12	0	-	0,8	<0,1	1
Luxembourg	-	-	-	-	-	-	-	-	-	-	-	-	-	-	-
Malta	Mediterranean Sea	Medium	173	<5%	7	0,04	6.414	100	399.867	100	-	-	0,1	<0,1	1
Poland	Baltic Sea	Low	634	30%	349	55	27.223	7	3.437.155	9	25	-	4,5	<0,1	3
Portugal	Atlantic Ocean	High	1187	<5%	338	28	122.082	72	8.379.748	80	4,55	-	0,7	0,2	2
Romania	Black Sea	Medium	226	50%	101	44,5	7.912	5	1.085.563	5	0	-	1,1	<0,1	2
Slovakia	-	-	-	-	-	-	-	-	-	-	-	-	-	-	-
Slovenia	Mediterranean Sea	Medium	45	<5%	14	30	3.102	9	299.465	15	0	-	0,1	<0,1	1
Spain	Atlantic Ocean Mediterranean Sea	Atlantic Ocean: High Mediterranean Sea: Medium	6.583	<5%	757	11,5	418.026	-	2.286.6485	-	3,64	1,6	1,6	0,4	2

	Maritime basin	Expected sea level rise	Coastal length	Zone below 5m elevation	Coastline subject to erosion		GDP in 50 km zone		Population in 50 km zone		Change in built-up in areas vulnerable to coastal floods	People in the 100-year flood plain	People flooded	Damage costs	Exp. to protect against coastal flood risk
Sweden	Baltic Sea	Medium	13.567	0%	327	2	119.904	52	6.282.989	70	10,17		0,2		1
The Netherlands	North Sea	High	1.275	>85%	134	11	241.116	53	8.941.918	55	8,54	10,2	5	2,3	4
United Kingdom	Atlantic Ocean North Sea	Atlantic Ocean: High North Sea: High	17.381	10-15%	3.009	16	1.090.642	69	46.565.867	78	13,31	4,4	4,8	1,2	4

Annex D. Projections by EurEau on costs of compliance with DWD and UWWTD

EurEau (2018) estimates investment needs to reach compliance for a number of utilities in member states, looking forward for ten years. These are set out in the Table below. They include investments for other purposes as well.

Table A D.1. Investment needs in water infrastructure reported by member states

For the next 10 years unless otherwise stated – in EUROs

	DWD	UWWTD	Asset renewal	Other
Austria (i)	222m p.a. (for next 5 years)	436m p.a. (for 5 years)		
Cyprus (ii)	275.15m total	73.45m	121.10m for new challenges[1] 0.2m in agri pollution, energy production 13m for operation/process optimisation	
Denmark (iii)	1bn (includes asset renewal)	4bn (includes asset renewal)		Drinking Water: 1.2bn[1] Wastewater: 5.5bn[1]
Finland			320m p.a.[2]	
France	3.5bn p.a.		DW: 3 – 5.4bn p.a. WW: 2.5 – 4.5bn p.a. Drainage 400 – 700m p.a.	DW: 5m p.a.[1] WW: 1-2bn p.a.[1]
Germany				Total 75bn over 10 years
Hungary			4.7bn	239m for smart metering
Ireland	4.1bn	3.32bn	835m[1]	118m for operation/process optimisation
Italy	20bn		25bn	10bn[1] 3bn other sectors 7bn operation/process optimisation
Norway	2.2bn		7.5bn	2.3bn[1] 5bn other sectors 10% of total (1.9bn) for operation/process optimisation
Poland		2003-15 WWTP: 2bn Sewers: 10.3bn 2016-21 WWTP: 2.3bn Sewers: 4.67bn	8.4bn	
Romania	6.68bn	10.38bn		
Spain (ii)	24.56bn[3]		18.63bn[4]	2.85bn[1,5] 9.13bn other sectors (agriculture) 4.07bn operation/process optimisation[6]
Sweden	10bn (includes operation/process		8bn	1bn[1]

	DWD	UWWTD	Asset renewal	Other
	optimisation)			
The Netherlands	5.6bn		Water: 1.9bn Other: 3.3bn	1.8bn[1] 0.4bn other sectors 1.3bn operation/process optimisation

Note: (i) Flood protection is not included; €40m p.a. projected for ecological measures. (ii) includes flood protection. (iii) unclear if flooding is included.
1 includes micropollutants, climate change and security
2 currently Finland spends some €120bn p.a. repairing, renewing and replacing water and wastewater assets
3 €5.33bn of this has already been invested in water supply (60%) and sanitation (40%)
4 existing asset value €2000/ person assuming a useful life of 50 years, requiring investment rates of €40/person p.a.
5 water reuse costs €1.4bn; CSO costs €1.45bn
6 assuming 10% of the costs of programme of measures (€3.48bn) plus a further 10% from other national figures (€590m)
Source: EurEau (2018).

Table A D.2. Expenditure needs to renew existing infrastructure reported by member states

For the next 10 years unless otherwise stated – in EUROs

Country	Reinvestment in aged infrastructure (EUR billon)	Total (reinvestment) EUR/cap/yr	Asset renewal as % of investment in compliance	Notes
Austria	2.18	49.70	196	5 year period (total investment was without flooding)
Cyprus	0.073	8.59	30	
Finland	3.2	58.15		
France	30-50	123.09	95 - 158	DWD
	25-45		79 - 142	UWWTD
	4 - 7		1 - 1.8	Urban drainage
Germany		22.64		
Hungary	4.7	48.03		
Ireland	3.32	69.51	90	
Italy	25	41.26	139	
Norway	7.5	142.63	379	
Poland	8.4	22.09	134	
Romania		35.08		Assume sum of 'investment values for water supply 2014 - 2020' and 'investment values for wastewater collection and treatment 2014 -2020'
Spain	18.63	40.04	82	total included 8% for flooding
Sweden	8	80.04	89	total included costs for operation and optimisation
The Netherlands	1.9	30.44	38	DWD
	3.3		65	UWWTD
Average			114	

Source: EurEau 2017, 2018.

Annex E. Assessment of RBIs and RFIs to finance flood protection

Table A E.1. Assessment of RBIS and RFIs to finance flood protection

Instrument	Land taxes	Earmarking of water abstraction and water body use charge	Offset schemes	PES schemes
Real-life examples presented in this report	Taxes for dyke maintenance (NL), Taxes for landowners protected by a dyke (DE)	Proposal in the Po RBD (IT)	Drinking water forests (DE)	Bosco Limite project (IT), SCaMP programme (UK)
1. Revenue-raising capacity	NL: revenues cover up to 95% of the costs of water quality and flood protection activities (Filatova, 2014). DE: no information could be found. In general, this instrument can raise significant revenues for flood protection, if tax levels are properly set.	As at the moment is just a preliminary proposal, it is not possible to estimate the revenue-raising potential.	Offset schemes offer a good opportunity to integrate nature protection funding with private capital. As these are voluntary schemes, however, it is likely that offset schemes will never be anything more than an additional source to be combined with other funds.	PES schemes offer opportunities to integrate nature protection funding with private capital. As these are voluntary schemes, however, it is likely that they will never be anything more than an additional source to be combined with other funds.
2. Capacity to steer behaviour	NL: every stakeholder pays accordingly to their interests – so, in principle, stakeholder more exposed pay more, and this provides an incentive for locating activities in areas with lower flood risk. DE: no information could be found. Overall, if tax rates are set in relation to exposure levels, this tax provides an incentive for locating activities, constructing of buying buildings outside at-risk areas.	It is just a proposal, so any considerations on this topic is purely preliminary. Earmarking water use and water body use fees for financing flood protection might not convey the right message to users, as users do not necessarily increase vulnerability. In contrast, a land use tax on flood-prone areas would be more effective in providing an incentive for risk-reduction behaviour, as building in areas not at risk would then become more convenient than building in risk-prone areas.	Not applicable – private companies compensate for their environmental impact by restoring ecosystems elsewhere. This does not necessarily imply that they will also reduce the impact caused by their activity. The schemes are set for nature protection or ecosystem restoration projects; flood protection (e.g. runoff reduction) is a secondary benefit rather than the main reason for setting up the scheme. Further research could focus on whether offset schemes specific for flood risk management do exist.	As in the Bosco Limite project, often PES schemes reward landowners applying good practices on their land: in this sense, PES schemes are a way to reward behavioural change. In the case of flood protection, in fact, PES scheme can focus on risk mitigation behaviour such as for example reforestation or terracing. On the other hand, the scale of such behavioural change will depend on the scale at which the scheme is applied.

Instrument	Land taxes	Earmarking of water abstraction and water body use charge	Offset schemes	PES schemes
3. Adaptability of the financing source	A land tax is generally paid by home and landowners every year, and thus is ensures a constant revenue flow over the years. Thus, revenues from these taxes are well suited for financing modular expenditures in flood protection over time. However, as these taxes ensure quite a steady cash flow over the years, they allow for planning large investments, for example by setting aside. It can therefore be said that this instrument is quite adaptable.	The same considerations made for land taxes apply to this case. Water abstraction and water body use charge are paid by users every year, and this ensures a constant revenue flow over the years. This makes revenues from these charges well suited for financing modular expenditures over the years. At the same time, as they ensure quite a steady cash flow over the years, these charges allow for planning large investments, e.g. by setting aside. Thus, this instrument is quite adaptable.	The adaptability is low, as offset schemes generally focus on a specific protection or restoration action. In the German case, for example, the bottled water company focuses on reforestation, so it is one-shot intervention on each parcel.	PES schemes can be implemented to compensate for specific practices/ actions or for environmentally-sound management of agricultural land. Thus, they can be used to finance day-by-day management of one-shot small measures, such as for example reforestation of a plot of land. They cannot be used to finance large infrastructures.
4. Allocation of costs across public and private investors	Through this instrument, the costs of flood protection are charged on citizens and businesses located in at-risk areas – and thus costs are not born by public authorities. Nevertheless, public authorities are still in charge of the management of such revenue flow, including the transaction costs – and thus the public sector still bears some burden, whereas public investors are not involved at all.	Through these charges, water users would finance flood protection activities – even though the general principle might be arguable (see criterion 2). The costs (or part thereof) are no longer borne by public authorities. Public authorities would still be in charge of the management of revenue flows, including the transaction costs – and thus the public sector would still bear some burden, whereas public investors are not involved at all.	Offset schemes are a good opportunity to inject private funds into nature protection. However, usually the schemes are set for nature protection or ecosystem restoration projects, and flood protection (e.g. runoff reduction) is a secondary benefit rather than the main reason for setting up the scheme.	PES schemes are a good opportunity to inject private funds into nature protection. However, flood protection (e.g. runoff reduction) might be just one of the ecosystem services traded under the scheme, rather than the main reason. Also, the magnitude of private capital raised will depend on the ecosystem services provided by the scheme and on the geographical scale of the scheme.
5. Geographical scale, and possibilities to scale up	This instrument can work at the local, district, regional and/or national level, depending on the administrative structure of a country.	This instrument can work at the local, district, regional and/or national level, depending on the administrative structure of a country.	Offset schemes generally focus on a specific protection or restoration action: thus, even if the scheme might work at the national level, it focuses on localized measures and interventions. To date, to the author's knowledge, there are no large schemes coordinating offset payments at a large scale, but this possibility could be further investigated.	PES schemes often focus on specific measures or management practices, as shown by the SCaMP example. It is unlikely that they can be used for financing natural flood management as a whole at the watershed level; they will rather be part of a wider financing strategy. In addition, they are often implemented at local or regional scales. Nevertheless, they can be scaled up at watershed scale. Careful planning might identify the scope for setting up multiple schemes at multiple locations (no example is available to the author's knowledge).

Instrument	Land taxes	Earmarking of water abstraction and water body use charge	Offset schemes	PES schemes
6. Replicability in other countries	This instrument can be adapted to the administrative structures of almost any country – especially considering that most country have some land tax system in place, even though it might not be linked with flood protection investments.	This instrument can be adapted to the administrative structures of almost any country – especially considering that most country have water abstraction charges in place.	In the German case, the offset scheme is an initiative of a private company and a nature protection association, working in agreement with public authorities and private landowners. In principle, in such a form, there should be no inconvenient for implementing a similar scheme elsewhere.	PES schemes are often based on private agreements between ecosystem services providers and buyers (public and private). The SCaMP programme was initiated by a water utility, and is implemented in association with an environmental NGO and the public forestry department, and the collaboration of OFWAT. The Bosco project bridges together private and public actors. European governance structures might pose some constraints to the development of PES schemes. But PES voluntary nature makes it possible to build agreements outside existing governance structures or in association with such structures.

Source: Acteon (2018), *Investment Needs and Innovative Financing Mechanisms for Flood Protection*, unpublished report to the OECD, Pari

www.ingramcontent.com/pod-product-compliance
Ingram Content Group UK Ltd.
Pitfield, Milton Keynes, MK11 3LW, UK
UKHW050413240426
12048UKWH00020B/1489